Surveys and Tutorials in the Applied Mathematical Sciences

Volume 8

Featuring short books of approximately 80-200pp, Surveys and Tutorials in the Applied Mathematical Sciences (STAMS) focuses on emerging topics, with an emphasis on emerging mathematical and computational techniques that are proving relevant in the physical, biological sciences and social sciences. STAMS also includes expository texts describing innovative applications or recent developments in more classical mathematical and computational methods.

This series is aimed at graduate students and researchers across the mathematical sciences. Contributions are intended to be accessible to a broad audience, featuring clear exposition, a lively tutorial style, and pointers to the literature for further study. In some cases a volume can serve as a preliminary version of a fuller and more comprehensive book.

More information about this series at http://www.springer.com/series/7219

Houman Owhadi · Clint Scovel · Gene Ryan Yoo

Kernel Mode Decomposition and the Programming of Kernels

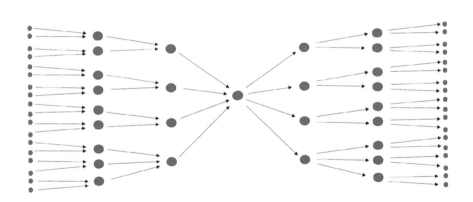

 Springer

Houman Owhadi (iD)
Computing and Mathematical Sciences
California Institute of Technology
Pasadena, CA, USA

Clint Scovel
Computing and Mathematical Sciences
California Institute of Technology
Pasadena, CA, USA

Gene Ryan Yoo (iD)
Quantitative Research
Susquehanna International Group
Bala Cynwyd, PA, USA

ISSN 2199-4765   ISSN 2199-4773   (electronic)
Surveys and Tutorials in the Applied Mathematical Sciences
ISBN 978-3-030-82170-8   ISBN 978-3-030-82171-5   (eBook)
https://doi.org/10.1007/978-3-030-82171-5

Mathematics Subject Classification: 62J02, 68T01, 68T10, 42C15

This Springer imprint is published by the registered company Springer Nature Switzerland AG
The registered company address is: Gewerbestrasse 11, 6330 Cham, Switzerland

Preface

There is currently a growing interest in using learning techniques to solve numerical approximation problems. These learning techniques can be divided into two categories: (1) The first category, composed of Kernel methods, stems from interplays between Gaussian process regression and numerical approximation and the fact that these two are intimately connected through the process of making estimations with partial information. (2) The second category, composed of variants of artificial neural networks, stems from the popularity of deep learning and the popularization of automatic differentiation in Python. Although the first category has strong theoretical foundations, it relies on the prior selection of a kernel, and our understanding of this selection problem is not fully developed. Although the second approach does not have much theoretical support yet, it can also be interpreted as a kernel method in which the kernel (parameterized by the inner layers of the network, and defined by the feature map identified as the output of those inner layers) is learned from data. Using a generalization of Huang's empirical mode decomposition problem as an overarching running example/application, one purpose of this book is to show how this kernel selection/learning problem can be addressed from the perspective of programming kernels. The programming of these kernels is achieved (through the construction of interpretable regression networks) by assembling elementary modules decomposing and recomposing kernels and data. These elementary steps are repeated across levels of abstraction and interpreted from the equivalent perspectives of optimal recovery, game theory, and Gaussian process regression (GPR). The prototypical mode/kernel decomposition module produces an approximation $(w_1, w_2, \cdots, w_m)$ of an element $(v_1, v_2, \ldots, v_m) \in V_1 \times \cdots \times V_m$ of a product of Hilbert subspaces $(V_i, \| \cdot \|_{V_i})$ of a common Hilbert space from the observation of the sum $v := v_1 + \cdots + v_m \in V_1 + \cdots + V_m$. This approximation is minmax optimal with respect to the relative error in the product norm $\sum_{i=1}^{m} \| \cdot \|_{V_i}^2$ and obtained as $w_i = Q_i (\sum_j Q_j)^{-1} v = \mathbb{E}[\xi_i | \sum_j \xi_j = v]$ where Q_i and $\xi_i \sim \mathcal{N}(0, Q_i)$ are the covariance operator and the Gaussian process defined by the norm $\| \cdot \|_{V_i}$. The prototypical mode/kernel recomposition module performs partial sums of the recovered modes w_i and covariance operators Q_i based on the alignment between each recovered mode w_i and the data v with respect to the inner product defined by S^{-1} with $S := \sum_i Q_i$ (which has a natural interpretation as model/data alignment $\langle w_i, v \rangle_{S^{-1}} = \mathbb{E}[\langle \xi_i, v \rangle_{S^{-1}}^2]$ and variance decomposition in the GPR setting). We illustrate the proposed framework by programming regression networks approximating the modes $v_i = a_i(t)y_i\big(\theta_i(t)\big)$ of a (possibly noisy) signal $\sum_i v_i$ when the amplitudes a_i, instantaneous phases θ_i and periodic waveforms y_i may all be unknown and show near machine precision recovery under regularity and separation assumptions on the instantaneous amplitudes a_i and frequencies $\dot{\theta}_i$. Our presentation includes a review of generalized additive models [45, 46], additive kernels/Gaussian processes [30, 87]; generalized Tikhonov regularization [31]; empirical mode decomposition [58]; and Synchrosqueezing [22], which are all closely related to and generalizable under the proposed framework. Python source codes are available at https://github.com/kernel-enthusiasts/Kernel-Mode-Decomposition-1D.

Pasadena, CA, USA   Houman Owhadi

Pasadena, CA, USA   Clint Scovel

Bala Cynwyd, PA, USA   Gene Ryan Yoo

Acknowledgments

The authors gratefully acknowledge support by the Air Force Office of Scientific Research under award number FA9550-18-1-0271 (Games for Computation and Learning). The authors also thank Peyman Tavallali and an anonymous referee for a careful read of the manuscript and helpful suggestions.

Contents

Chapter 1

Introduction

1.1 The Empirical Mode Decomposition Problem

The purpose of the *Empirical Mode Decomposition* (EMD) algorithm [58] can be loosely expressed as solving a (usually noiseless) version of the following problem, illustrated in Fig. 1.1.

Problem 1. *For $m \in \mathbb{N}^*$, let $a_1, \ldots, a_m$ be piecewise smooth functions on $[0, 1]$, and let $\theta_1, \ldots, \theta_m$ be strictly increasing functions on $[0, 1]$. Assume that m and the a_i and θ_i are unknown. Given the (possibly noisy) observation of $v(t) = \sum_{i=1}^m a_i(t) \cos\left(\theta_i(t)\right), t \in [0, 1]$, recover the modes $v_i(t) := a_i(t) \cos\left(\theta_i(t)\right)$.*

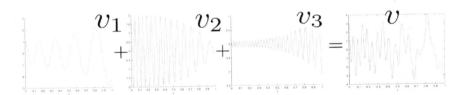

Figure 1.1: A prototypical mode decomposition problem: given $v = v_1 + v_2 + v_3$ recover v_1, v_2, and v_3.

In practical applications, generally the *instantaneous frequencies* $\omega_i = \frac{d\theta_i}{dt}$ are assumed to be smooth and well separated. Furthermore, the ω_i and the *instantaneous amplitudes* are assumed to be varying at a slower rate than the *instantaneous phases* θ_i so that near $\tau \in [0, 1]$ the *intrinsic mode function* v_i can be approximated by a trigonometric function, i.e.,

$$v_i(t) \approx a_i(\tau) \cos\left(\omega_i(\tau)(t - \tau) + \theta_i(\tau)\right) \text{ for } t \approx \tau. \qquad (1.1.1)$$

© The Author(s), under exclusive license to Springer Nature Switzerland AG 2021
H. Owhadi et al., *Kernel Mode Decomposition and the Programming of Kernels*, Surveys and Tutorials in the Applied Mathematical Sciences 8, https://doi.org/10.1007/978-3-030-82171-5_1

The difficulty of analyzing and generalizing the EMD approach and its popularity and success in practical applications [57] have stimulated the design of alternative methods aimed at solving Problem 1. Methods that are amenable to a greater degree of analysis include synchrosqueezing [21, 67], variational mode decomposition [24], and nonlinear L_1 minimization with sparse time–frequency representations [53, 54].

1.2 Programming Kernels Through Regression Networks and Kernel Mode Decomposition

Mode Decomposition as a Rosetta Stone for Pattern Recognition Problem 1 is a prototypical pattern recognition problem that can be addressed from the perspectives of numerical approximation, statistical inference, and machine learning. One could therefore use its analysis, from the combined approaches of numerical approximation and statistical inference, as a Rosetta stone for developing our understanding of learning techniques such as artificial neural network (ANN) methods [40] and kernel methods [98].

Although successful industrial applications [65] have consolidated the recognition of ANNs as powerful pattern recognition tools, their utilization has recently been compared to "operating on an alien technology" [60] due to the challenges brought by a lag in theoretical understanding. Although kernel methods have strong theoretical foundations [98, 102, 107], they require the prior selection of a *good kernel*, and our understanding of this kernel selection problem is still in its infancy.

Learning Approach to the Kernel Selection Problem One natural approach to the kernel selection problem is to (1) start with a parameterized family of kernels and (2) learn the parameters of the kernels using Cross-Validation, Maximum Likelihood Estimation, or Maximum a Posteriori Estimation (MAP) [6, 13, 81, 84, 106]. In the setting of residual neural networks, [47] and [81] show that ANNs can also be interpreted and analyzed as kernel methods with data-dependent kernels (learned via MAP estimation if training is performed with L^2 regularization on weights and biases) parameterized by the inner layers of the network and defined by interpreting the output of the inner layers as a (parameterized) feature map. In the setting of learning dynamical systems, [44] shows that the accuracy of kernel methods improves by several orders of magnitude after learning from data (using a variant of cross-validation known as Kernel Flows [84]), and these kernel methods (with learned kernels) can outperform ANN-based methods in terms of both complexity and accuracy for the extrapolation of weather/climate series [42].

Programming Kernels Through Programmable and Interpretable Regression Networks One objective of this monograph is to present an alternative (programming) approach to the kernel selection problem while using mode decomposition [58] as a prototypical pattern recognition problem. In this approach, kernels are programmed for the task at hand based on a rational and modular (object-oriented) framework. This framework is based on the construction of interpretable Gaussian process regression (GPR)-based networks that are (1) programmable based on rational and modular (object-oriented) design and (2) amenable to analysis and convergence results. Since elementary operations performed by ANNs can be interpreted [86] as stacking Gaussian process regression steps with nonlinear thresholding and pooling operations across levels of abstractions, such networks may also help our understanding of fundamental mechanisms that might be at play in ANN-based pattern recognition.

On Numerical Approximation, Kernel Methods, and the Automation of the Process of Discovery Most numerical approximation methods can be interpreted as kernel interpolation methods [83], and the particular choice of the kernel can have a large impact on the accuracy/complexity of the method [43]. While the classical numerical approximation approach to kernel selection has been to use prior information on the regularity of the target function or on the PDE solved by the target function [12, 82], such information is not always available, and one has to learn the kernel from the available data [13, 84]. Although regressors obtained from artificial neural networks can be interpreted as kernel regressors with warping kernels learned from data [81], they rely on trial and error for the identification of the architecture of the network (or equivalently, the parameterized family of kernels). Another objective of this monograph is to show how this process of trial and error can, to some degree, be replaced by modular programming techniques in the setting of additive kernels and the kernel mode decomposition methodology (introduced and advocated for here).

1.3 Mode Decomposition with Unknown Waveforms

As an application of the programmable and interpretable regression networks introduced in this book, we will also address the following generalization of Problem 1, where the periodic waveforms may all be non-trigonometric, distinct, and unknown and present an algorithm producing near machine precision (10^{-7} to 10^{-4}) recoveries of the modes.

Problem 2. *For $m \in \mathbb{N}^*$, let $a_1, \ldots, a_m$ be piecewise smooth functions on $[-1, 1]$, let $\theta_1, \ldots, \theta_m$ be piecewise smooth functions on $[-1, 1]$ such that the instantaneous frequencies $\dot{\theta}_i$ are strictly positive and well separated, and let $y_1, \ldots, y_m$ be square-integrable 2π-periodic functions. Assume that m and the $a_i, \theta_i,$ and y_i are all unknown. Given the observation $v(t) = \sum_{i=1}^{m} a_i(t) y_i(\theta_i(t))$ (for $t \in [-1, 1]$), recover the modes $v_i(t) := a_i(t) y_i(\theta_i(t))$.*

One fundamental idea is that although Problems 1 and 2 are nonlinear, they can be, to some degree, linearized by recovering the modes v_i as aggregates of sufficiently fine modes living in linear spaces (which, as suggested by the approximation (1.1.1), can be chosen as linear spans of functions $t \to \cos(\omega(t-\tau)+\theta)$ windowed around τ, i.e., Gabor wavelets). The first part of the resulting network recovers those finer modes through a linear optimal recovery operation. Its second part recovers the modes v_i through a hierarchy of (linear) aggregation steps sandwiched between (nonlinear) ancestor/descendant identification steps. These identification steps are obtained by composing the alignments between v and the aggregates of the fine modes with simple and interpretable nonlinearities (such as thresholding, graph-cuts, etc.), as presented in Chap. 4.

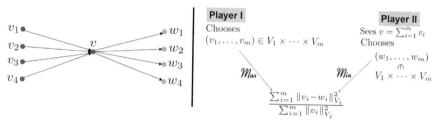

Figure 1.2: Left: the mode decomposition problem. Right: the game theoretic approach.

1.4   Structure of the Monograph

The structure of the monograph is as follows. Chapter 2 presents a review of additive Gaussian process regression, empirical mode decomposition, and synchrosqueezing, which are all related to or can be generalized in the kernel mode decomposition framework introduced here. Chapter 3 introduces a general (abstract) formulation of the *mode decomposition problem* and develops its solution in the context of the three separate fields of optimal recovery, game/decision theory, and Gaussian process regression. This chapter also includes a review of additive kernels and an illustrative solution to Problem 1 when only the amplitudes $a_i(t)$ are unknown (based on the prior knowledge of the number of quasi-periodic modes and their phase functions θ_i). Chapter 4 introduces kernel mode decomposition networks (KMDNets) as a development of Chap. 3. This chapter includes an illustrative solution to the full Problem 1. The following chapters illustrate the proposed approach by programming kernels (via KMDNets) for increasingly complex (mode decomposition) tasks. Consequently, Chap. 5 continues the development of KMDNets by introducing additional programming modules and illustrates their design by developing flexible versions of synchrosqueezing, due to Daubechies et al. [21, 22]. This chapter includes an illustrative solution to the full Problem 1 with crossing instantaneous frequencies of the modes. Chapter 6 considers the mode recovery Problem 1 generalized to the case where the base waveform of each mode is the same known, possibly non-trigonometric, square-integrable 2π-periodic function $t \to y(t)$ and develops an *iterated micro-local Kernel Mode Decomposition* algorithm for its solution. Chapter 7 considers the extension, Problem 2, of the mode recovery problem, Problem 1, to the case where the periodic base waveform of each mode is unknown and may be different across modes. Chapter 8 addresses a generalization of the mode recovery Problem 2 (with known waveforms), allowing for crossing frequencies, vanishing modes, and noise. The appendix describes the universality of the aggregated kernel (discussed in Sect. 5.3) and deferred proofs.

Chapter 2

Review

The kernel mode decomposition framework has relations to the fields of additive Gaussian process regression, empirical mode decomposition, and synchrosqueezing. Consequently, here we review these subjects giving context to our work. This chapter is not essential to understanding the book and so can be skipped on first reading.

Although simple kriging and GPR are derived differently, they can be shown to be equivalent and are often referred to as the same, see e.g., Yoo [124, Sec. 1.1] for a review of kriging and its relationship with GPR. Regarding the origins of kriging, paraphrasing Cressie [16], known for introducing kriging in spatial statistics, "both Matheron [74] (see also [73]) and Gandin [38] were the first to publish a definitive development of spatial kriging." D. G. Krige's contributions in mining engineering were considerable, but he did not discover kriging, illustrating once again Stigler's Law of Eponymy (Stigler [108]), which states that "no scientific discovery is named after its original discoverer." The eponymous title of Stigler's work is playfully consistent with his law, since in it, he essentially names Merton [77, p. 356] as the discoverer of Stigler's law.

2.1 Additive Gaussian Processes

Following Hastie and Tibshirani [45, 46], the generalized additive model (GAM) replaces a linear predictor $\sum_j \beta_j x_j$, where the β_j are parameters, with $\sum_j f_j(x_j)$, where the f_j are unspecified functions. For certain types of prediction problems such as binary target variables, one may add a final function $h\left(\sum_j f_j(x_j)\right)$. To incorporate fully dependent responses, we can consider models of the form $f(x_1, \ldots, x_N)$. Additive models have been successfully used in regression, see Stone [109] and Fan et al. [32]. Vector-valued generalizations of GAMs have been developed in Yee and Wild [123] and Yee [122]. For vector-valued additive

H. Owhadi et al., *Kernel Mode Decomposition and the Programming of Kernels*, Surveys and Tutorials in the Applied Mathematical Sciences 8, https://doi.org/10.1007/978-3-030-82171-5_2

models of large vector dimension with a large number of dimensions in the observation data, Yee [122] develops methods for reducing the rank of the systems used in their estimation.

When the underlying random variables are Gaussian, and we apply to regression, we naturally describe the model in terms of its covariance kernel $k(x_1, \ldots, x_N, x'_1, \ldots, x'_N)$ or as an additive model $\sum k_i(x_i, x'_i)$, where the kernel is an additive sum of kernels depending on lower dimensional variables. It is natural to generalize this setting to a covariance defined by a weighted sum over all orders d of dependency of weighted sums of kernels depending only on d D dimensional variables, where $N = Dd$. Of course, such general kernels are exponentially complex in the dimension N, so they are not very useful. Nearly simultaneously, Duvenaud et al. [30] and Durrande et al. [26, 27], introducing Gaussian Additive Processes, addressed this problem. Duvenaud et al. [30] restrict the sum at order d to be symmetric in the scalar components in the vector variables and thus reduce this complexity in such a way that their complexity is mild and their estimation is computationally tractable. Durrande et al. [26, 27] consider additive versions of vector-dependent kernels and product versions of them and study their respective performance properties along with the performance of their sum. Moreover, because of the additive nature of these methodologies, they both achieve strong interpretability as described by Plate [87].

2.2 Gaussian Process Regression

Williams and Rasmussen [118] provide an introduction to Gaussian Process Regression (GPR). More generally, an excellent introduction to Gaussian processes in machine learning, along with a description of many of its applications and its history, can be found in Rasmussen [93] and Rasmussen and Williams [94], and see also Yoo [124]. Recent application domain developments include source separation, which is related to the subject of this book, by Park and Choi [85] and Liutkus et al. [69] and the detection of periodicities by Durrande et al. [28, 29], and Preoţiuc-Pietro and Cohn [88].

When the number of dimensions of the observational data is large, computational efficiency becomes extremely important. There has been much work in this area, the so-called *sparse methods*, e.g., Tresp [115], Smola and Bartlett [103], Williams and Seeger [117], Csató and Opper [18], Csató et al. [19], Csató [17], Quiñonero-Candela [89], Lawrence et al. [63], Seeger [100], Seeger et al. [101], Schwaighofer and Tresp [99], and Snelson and Ghahramani [104]. Quiñonero-Candela and Rasmussen [90] provide a unifying framework for the sparse methods based on expressing them in terms of their *effective prior*. The majority of these methods utilize the so-called *inducing variable* methods, which are data points in the same domain as the unlabeled data. Some require these to be a subset of the training data, while others, such as Snelson and Ghahramani [104], allow them to inferred along with the primary hyperparameters using optimization. However, there are notable exceptions, such as

Hensman et al. [49] who apply a Kullback–Leibler derived variational formulation and utilize Bochner's theorem on positive definite functions to choose optimal features in the Fourier space.

The majority of these methods use the Kullback–Leibler (KL) criterion to select the induced points; see Rasmussen and Williams [94, Ch. 8] for a review. In particular, Seeger et al. [101] and Seeger [100], among others, utilize the KL criterion to optimize both the model hyperparameters and the inducing variables. However, they observe that the approximation of the marginal likelihood is sensitive to the choice of inducing variables, and therefore convergence of the method is problematic. Snelson and Ghahramani [104] attempt to resolve this problem by developing a KL formulation where the model hyperparameters and the inducing variables are jointly optimized. Nevertheless, since the inducing variables determine an approximate marginal likelihood, these methods can suffer from overfitting. Titsias' [112] breakthrough, a development of Csató and Opper [18] and Seeger [100], was the introduction of a KL variational framework where the model hyperparameters and the inducing variables are selected in such a way as to maximize a lower bound to the true marginal likelihood and thus are selected to minimize the KL distance between the sparse model and the true one. When the dimensions of the observational data are very large, Hensman et al. [51], utilizing recent advances in *stochastic variational inference* of Hoffman et al. [52] and Hensman et al. [50], appear to develop methods which scale well. Adam et al. [1] develop these results in the context additive GP applied to the source separation problem.

For vector Gaussian processes, one can proceed basically as in the scalar case, including the development of sparse methods, and however one needs to take care that the vector covariance structure is positive definite (see the review by Alvarez et al. [4]); see e.g., Yu et al. [126], Boyle and Frean [10, 11], Melkumyan and Ramos [76], Alvarez and Lawrence [2, 3], and Titsias and Lázaro-Gredilla [113]. Raissi et al. [91] develop methods to learn linear differential equations using GPs.

2.3 Empirical Mode Decomposition (EMD)

The definition of an *instantaneous frequency* of a signal $x(t)$ is normally accomplished through application of the Hilbert transform $\mathcal{H}$ defined by the principle value of the singular integral

$$\big(\mathcal{H}(x)\big)(t) := \frac{1}{\pi} PV \int_{\mathbb{R}} \frac{x(\tau)}{t - \tau} d\tau,$$

which, when it is well defined, determines the harmonic conjugate $y := \mathcal{H}(x)$ of $x(t)$ of a function

$$x(t) + iy(t) = a(t)e^{i\theta(t)},$$

which has an analytic extension to the upper complex half plane in t, allowing the derivative $\omega := \dot{\theta}$ the interpretation of an instantaneous frequency of

$$x(t)) = a(t)\cos(\theta(t)).$$

However, this definition is controversial, see e.g., Boashash [8] for a review, and possesses many difficulties, and the Empirical Mode Decomposition (EMD) algorithm was invented by Huang et al. [58] to circumvent them by decomposing a signal into a sum of *intrinsic mode functions* (IMFs), essentially functions whose number of local extrema and zero crossings are either equal or differ by 1 and such that the mean of the envelope of the local maxima and the local minima is 0, which are processed without difficulty by the Hilbert transform. See Huang [56] for a more comprehensive discussion. This combination of the EMD and the Hilbert transform, called the Hilbert–Huang transform, is used to decompose a signal into its fundamental AM–FM components. Following Rilling et al. [96], the EMD appears as follows: given a signal $x(t)$,

1. identify all local extrema of $x(t)$,

2. interpolate between the local minima (respectively, maxima) to obtain the envelope $e_{\min}(t)$ (respectively, $e_{\max}(t)$),

3. compute the mean $m(t) := \frac{e_{\min}(t) + e_{\max}(t)}{2}$,

4. extract the detail $d(t) := x(t) - m(t)$, and

5. iterate on the residual $m(t)$

The *sifting* process iterates steps (1) through (4) on the detail until it is close enough to zero mean. Then, the residual is computed and step (5) is applied.

Despite its remarkable success, see e.g., [14, 15, 20, 23, 58, 105, 121] and the review on geophysical applications of Huang and Wu [59], the original method is defined by an algorithm, and therefore its performance is difficult to analyze. In particular, sifting and other iterative methods usually do not allow for backward error propagation. Despite this, much is known about it, improvements have been made, and efforts are underway to develop formulations that facilitate a performance analysis. To begin, it appears that the EMD algorithm is sensitive to noise, so that Wu and Huang [120] introduce and study an Ensemble EMD, further developed in Torres et al. [114], which appears to resolve the noise problem while increasing the computational costs. On the other hand, when applied to white noise, Flandrin et al. [34–36] and Wu and Huang [119] demonstrate that it acts as an adaptive wavelet-like filter bank, leading to Gilles' [39] development of empirical wavelets. Rilling and Flandrin [95] successfully analyze the performance of the algorithm on the sum of two cosines. Lin et al. [68] consider an alternative framework for the empirical mode decomposition problem considering a moving average operator instead of the mean function of the EMD. This leads to a mathematically analyzable framework and in some cases (such as the stationary case) to the analysis of Toeplitz operators, a good theory with good results. This technique has been further developed by Huang et al. [55], with some success. Approaches based on variational principles, such as Feldman [33], utilizing an iterative variational approach using the Hilbert transform, Hou and Shi [53], a compressed sensing approach, Daubechies et al. [21], the wavelet base *synchrosqueezing* method to be discussed in a moment, and Dragomiretskiy and Zosso [24], a generalization of the classic Wiener filter using the alternate direction method of multipliers method, see Boyd et al. [9], to solve the resulting bivariate minimization problem, appear to be good candidates for analysis. However, the variational objective function in [53] uses higher order total variational terms so appears sensitive to noise, [33] is an iterative variational approach, and the selection of the relevant modes in [24] for problems with noise is currently under investigation, see e.g., Ma et al. [71] and the references therein. On the other hand, Daubechies et al. [21] provide rigorous performance guarantees under certain conditions. Nevertheless, there is still much effort in developing their work, see e.g., Auger et al. [5] for a review of synchrosqueezing and its relationship with time–frequency reassignment.

2.4   Synchrosqueezing

Synchrosqueezing, introduced in Daubechies and Maes [22], was developed in Daubechies, Lu and Wu [21] as an alternative to the EMD algorithm, which would allow mathematical performance analysis, and has generated much interest, see e.g., [5, 66, 79, 110, 111, 116]. Informally following [21], for a signal $x(t)$, we let

$$W(a,b) := a^{-\frac{1}{2}} \int_{\mathbb{R}} x(t) \overline{\psi\left(\frac{t-b}{a}\right)} dt$$

denote the wavelet transform of the signal $x(t)$ using the wavelet ψ. They demonstrate that for a wavelet such that its Fourier transform satisfies $\hat{\psi}(\xi) = 0, \xi < 0$, when applied to a pure tone

$$x(t) := A\cos(\omega t) \tag{2.4.1}$$

that

$$\omega(a,b) := -i\frac{\partial \ln W(a,b)}{\partial b} \tag{2.4.2}$$

satisfies

$$\omega(a,b) = \omega,$$

that is, it provides a perfect estimate of the frequency of the signal (2.4.1). This suggests using (2.4.2) to define the map

$$(a,b) \mapsto (\omega(a,b),b)$$

to push the mass in the reconstruction formula

$$x(b) = \Re\left[C_\psi^{-1} \int_0^\infty W(a,b)a^{-\frac{3}{2}} da\right],$$

where $C_\psi := \int_0^\infty \frac{\overline{\hat{\psi}(\xi)}}{\xi} d\xi$, to obtain the identity

$$Re\left[C_\psi^{-1} \int_0^\infty W(a,b)a^{-\frac{3}{2}} da\right] = \Re\left[C_\psi^{-1} \int_{\mathbb{R}} T(\omega,b)d\omega\right], \tag{2.4.3}$$

where

$$T(\omega, b) = \int_{A(b)} W(a, b) a^{-\frac{3}{2}} \delta\big(\omega(a, b) - \omega\big) da, \tag{2.4.4}$$

where

$$A(b) := \{a : W(a, b) \neq 0\}$$

and $\omega(a, b)$ is defined as in (2.4.2) for (a, b) such that $a \in A(b)$. We therefore obtain the reconstruction formula

$$x(b) = \Re\left[C_\psi^{-1} \int_{\mathbb{R}} T(\omega, b) d\omega \right] \tag{2.4.5}$$

for the synchrosqueezed transform T. In addition, [21, Thm. 3.3] demonstrates that for a signal x comprised of a sum of AM–FM modes with sufficiently separated frequencies whose amplitudes are slowly varying with respect to their phases and that the synchrosqueezed transform $T(\omega, b)$ is concentrated in narrow bands $\omega \approx \dot{\theta}_i(b)$ about the instantaneous frequency of the i-th mode and restricting the integration in (2.4.5) to these bands provides a good recovery of the modes.

Chapter 3

The Mode Decomposition Problem

In this chapter, we formulate a general (abstract) formulation of the *mode decomposition problem* and develop its solution in the context of the three separate fields of optimal recovery, game/decision theory, and Gaussian process regression. To that end, let V be a separable Hilbert space with inner product $\langle \cdot, \cdot \rangle$ and corresponding norm $\| \cdot \|$. Also let $\mathcal{I}$ be a finite set of indices and let $(V_i)_{i \in \mathcal{I}}$ be linear subspaces $V_i \subset V$ such that

$$V = \sum_{i \in \mathcal{I}} V_i \,. \tag{3.0.1}$$

The mode decomposition problem can be informally formulated as follows

Problem 3. *Given $v \in V$ recover $v_i \in V_i, i \in \mathcal{I}$, such that $v = \sum_{i \in \mathcal{I}} v_i$.*

Our solution to Problem 3 will use the interface between numerical approximation, inference, and learning (as presented in [82, 83]), which, although traditionally seen as entirely separate subjects, are intimately connected through the common purpose of making estimations with partial information [83]. Since the study of this interface has been shown to help automate the process of discovery in numerical analysis and the design of fast solvers [80, 82, 97], this book is also motivated by the idea it might, in a similar manner and to some degree, also help the process of discovery in machine learning. Here, these interplays will be exploited to address the general formulation Problem 3 of the mode recovery problem from the three perspectives of optimal recovery, game theory, and Gaussian process regression. The corresponding minimax recovery framework (illustrated in Fig. 1.2 and presented below) will then be used as a building block for the proposed programmable networks.

H. Owhadi et al., *Kernel Mode Decomposition and the Programming of Kernels*, Surveys and Tutorials in the Applied Mathematical Sciences 8, https://doi.org/10.1007/978-3-030-82171-5_3

3.1   Optimal Recovery Setting

Problem 3 is ill-posed if the subspaces $(V_i)_{i \in \mathcal{I}}$ are not linearly independent, in the sense that such a recovery will not be unique. Nevertheless, optimal solutions can be defined in the optimal recovery setting of Micchelli, and Rivlin [78]. To this end, let $\| \cdot \|_\mathcal{B}$ be a quadratic norm on the product space

$$\mathcal{B} = \prod_{i \in \mathcal{I}} V_i, \tag{3.1.1}$$

making $\mathcal{B}$ a Hilbert space, and let

$$\Phi : \mathcal{B} \to V$$

be the information map defined by

$$\Phi u := \sum_{i \in \mathcal{I}} u_i, \qquad u = (u_i)_{i \in \mathcal{I}} \in \mathcal{B}. \tag{3.1.2}$$

An optimal recovery solution mapping

$$\Psi : V \to \mathcal{B}$$

for the mode decomposition problem is defined as follows: for given $v \in V$, we define $\Psi(v)$ to be the minimizer w of

$$\min_{w \in \mathcal{B} | \Phi w = v} \max_{u \in \mathcal{B} | \Phi u = v} \frac{\|u - w\|_\mathcal{B}}{\|u\|_\mathcal{B}}. \tag{3.1.3}$$

Lemma 3.1.1. *Let* $\Phi : \mathcal{B} \to V$ *be surjective. For* $v \in V$, *the solution* w *of the convex optimization problem*

$$\begin{cases} \textit{Minimize } \|w\|_\mathcal{B} \\ \textit{Subject to } w \in \mathcal{B} \textit{ and } \Phi w = v \end{cases} \tag{3.1.4}$$

determines the unique optimal minmax solution $w = \Psi(v)$ *to* (3.1.3). *Moreover,*

$$\Psi(v) = \Phi^+ v,$$

where the Moore-Penrose inverse $\Phi^+ : V \to \mathcal{B}$ *of* Φ *is defined by*

$$\Phi^+ := \Phi^T (\Phi \Phi^T)^{-1}.$$

Now let us be more specific about the structure of $\mathcal{B}$ that we will assume. Indeed, let the subspaces $(V_i)_{i \in \mathcal{I}}$ be equipped with quadratic norms $(\| \cdot \|_{V_i})_{i \in \mathcal{I}}$ making each

$$(V_i, \| \cdot \|_{V_i})$$

a Hilbert space, and equip their product $\mathcal{B} = \prod_{i \in \mathcal{I}} V_i$ with the product norm

$$\|u\|_\mathcal{B}^2 := \sum_{i \in \mathcal{I}} \|u_i\|_{V_i}^2, \qquad u = (u_i)_{i \in \mathcal{I}} \in \mathcal{B}. \tag{3.1.5}$$

We use the notation $[\cdot, \cdot]$ for the duality product between V^* on the left and V on the right, and also for the duality product between V_i^* and V_i for all i. The norm $\| \cdot \|_{V_i}$ makes V_i into a Hilbert space if and only if

$$\|v_i\|_{V_i}^2 = [Q_i^{-1}v_i, v_i], \qquad v_i \in V_i, \tag{3.1.6}$$

for some positive symmetric linear bijection

$$Q_i : V_i^* \to V_i,$$

where by positive and symmetric we mean $[\phi, Q_i\phi] \geqslant 0$ and $[\phi, Q_i\varphi] = [\varphi, Q_i\phi]$ for all $\varphi, \phi \in V_i^*$. For each $i \in \mathcal{I}$, the dual space V_i^* to $(V_i, \| \cdot \|_{V_i})$ is also a Hilbert space with norm

$$\|\phi_i\|_{V_i^*}^2 := [\phi_i, Q_i\phi_i], \qquad \phi_i \in V_i^*, \tag{3.1.7}$$

and therefore the dual space $\mathcal{B}^*$ of $\mathcal{B}$ can be identified with the product of the dual spaces

$$\mathcal{B}^* = \prod_{i \in \mathcal{I}} V_i^* \tag{3.1.8}$$

with (product) duality product

$$[\phi, u] = \sum_{i \in \mathcal{I}} [\phi_i, u_i], \qquad \phi = (\phi_i)_{i \in \mathcal{I}} \in \mathcal{B}^*, \quad u = (u_i)_{i \in \mathcal{I}} \in \mathcal{B}. \tag{3.1.9}$$

Moreover the symmetric positive linear bijection

$$Q : \mathcal{B}^* \to \mathcal{B} \tag{3.1.10}$$

defining the quadratic norm $\| \cdot \|_{\mathcal{B}}$ is the block-diagonal operator

$$Q := \mathrm{diag}(Q_i)_{i \in \mathcal{I}}$$

defined by its action $Q\phi = (Q_i\phi_i)_{i \in \mathcal{I}}$, $\phi \in \mathcal{B}^*$.

Let

$$e_i : V_i \to V$$

be the subset inclusion and let its adjoint

$$e_i^* : V^* \to V_i^*$$

be defined through $[e_i^*\phi, v_i] = [\phi, e_i v_i]$ for $\phi \in V^*, v_i \in V_i$. These operations naturally transform the family of operators

$$Q_i : V_i^* \to V_i, \quad i \in \mathcal{I},$$

into a family of operators

$$e_i Q_i e_i^* : V^* \to V, \quad i \in \mathcal{I},$$

all defined on the same space, so that we can define their sum $S : V^* \to V$ by

$$S = \sum_{i \in \mathcal{I}} e_i Q_i e_i^* . \tag{3.1.11}$$

The following proposition demonstrates that S is invertible and that S^{-1} and S naturally generate dual Hilbert space norms on V and V^*, respectively.

Lemma 3.1.2. *The operator* $S : V^* \to V$, *defined in* (3.1.11), *is invertible. Moreover,*

$$\|v\|_{S^{-1}}^2 := [S^{-1}v, v], \quad v \in V, \tag{3.1.12}$$

defines a Hilbert space norm on V *and*

$$\|\phi\|_S^2 := [\phi, S\phi] = \sum_{i \in \mathcal{I}} \|e_i^* \phi\|_{V_i^*}^2, \quad \phi \in V^*, \tag{3.1.13}$$

defines a Hilbert space norm on V^* *which is dual to that on* V.

The following theorem determines the optimal recovery map Ψ.

Theorem 3.1.3. *For* $v \in V$, *the minimizer of* (3.1.4) *and therefore the min-max solution of* (3.1.3) *is*

$$\Psi(v) = \left(Q_i e_i^* S^{-1} v \right)_{i \in \mathcal{I}} . \tag{3.1.14}$$

Furthermore

$$\Phi\big(\Psi(v)\big) = v, \quad v \in V,$$

where

$$\Psi : (V, \|\cdot\|_{S^{-1}}) \to (\mathcal{B}, \|\cdot\|_{\mathcal{B}})$$

and

$$\Phi^* : (V^*, \|\cdot\|_S) \to (\mathcal{B}^*, \|\cdot\|_{\mathcal{B}^*})$$

are isometries. In particular, writing $\Psi_i(v) := Q_i e_i^* S^{-1} v$, *we have*

$$\|v\|_{S^{-1}}^2 = \|\Psi(v)\|_{\mathcal{B}}^2 = \sum_{i \in \mathcal{I}} \|\Psi_i(v)\|_{V_i}^2 \quad v \in V . \tag{3.1.15}$$

Observe that the adjoint

$$\Phi^* : V^* \to \mathcal{B}^*$$

of $\Phi : \mathcal{B} \to V$, defined by $[\varphi, \Phi u] = [\Phi^*(\varphi), u]$ for $\varphi \in V^*$ and $u \in \mathcal{B}$, is computed to be

$$\Phi^*(\varphi) = (e_i^* \varphi)_{i \in \mathcal{I}}, \quad \varphi \in V^* . \tag{3.1.16}$$

The following theorem presents optimality results in terms of Φ^*.

Theorem 3.1.4. *We have*

$$\|u - \Psi(\Phi u)\|_{\mathcal{B}}^2 = \inf_{\phi \in V^*} \|u - Q\Phi^*(\phi)\|_{\mathcal{B}}^2 = \inf_{\phi \in V^*} \sum_{i \in \mathcal{I}} \|u_i - Q_i e_i^* \phi\|_{V_i}^2 . \tag{3.1.17}$$

3.2 Game/Decision Theoretic Setting

Optimal solutions to Problem 3 can also be defined in the setting of the game/decision theoretic approach to numerical approximation presented in [82]. In this setting the minmax problem (3.1.3) is interpreted as an adversarial zero sum game (illustrated in Fig. 1.2) between two players and lifted to mixed strategies to identify a saddle point. Let $\mathcal{P}_2(\mathcal{B})$ be the set of Borel probability measures μ on $\mathcal{B}$ such that $\mathbb{E}_{u \sim \mu}\big[\|u\|_{\mathcal{B}}^2\big] < \infty$, and let $L(V, \mathcal{B})$ be the set of Borel measurable functions $\psi : V \to \mathcal{B}$. Let $\mathcal{E} : \mathcal{P}_2(\mathcal{B}) \times L(V, \mathcal{B}) \to \mathbb{R}$ be the loss function defined by

$$\mathcal{E}(\mu, \psi) = \frac{\mathbb{E}_{u \sim \mu}\big[\|u - \psi(\Phi u)\|_{\mathcal{B}}^2\big]}{\mathbb{E}_{u \sim \mu}\big[\|u\|_{\mathcal{B}}^2\big]}, \qquad \mu \in \mathcal{P}_2(\mathcal{B}), \psi \in L(V, \mathcal{B}). \qquad (3.2.1)$$

Let us also recall the more general notion of a Gaussian field as described in [82, Chap. 17]. To that end, a Gaussian space $\mathbf{H}$ is a linear subspace $\mathbf{H} \subset L^2(\Omega, \Sigma, \mathbb{P})$ of the L^2 space of a probability space consisting of centered Gaussian random variables. A centered Gaussian field ξ on $\mathcal{B}$ with covariance operator $Q : \mathcal{B}^* \to \mathcal{B}$, written $\xi \sim \mathcal{N}(0, Q)$, is an isometry

$$\xi : \mathcal{B}^* \to \mathbf{H}$$

from $\mathcal{B}^*$ to a Gaussian space $\mathbf{H}$, in that

$$[\phi, \xi] \sim \mathcal{N}\big(0, [\phi, Q\phi]\big), \qquad \phi \in \mathcal{B}^*,$$

where we use the notation $[\phi, \xi]$ to denote the action $\xi(\phi)$ of ξ on the element $\phi \in \mathcal{B}^*$, thus indicating that ξ is a weak $\mathcal{B}$-valued Gaussian random variable. As discussed in [82, Chap. 17], there is a one to one correspondence between Gaussian cylinder measures and Gaussian fields.[1] Let ξ denote the Gaussian field

$$\xi \sim \mathcal{N}(0, Q)$$

on $\mathcal{B}$ where $Q : \mathcal{B}^* \to \mathcal{B}$ is the block-diagonal operator $Q := \mathrm{diag}(Q_i)_{i \in \mathcal{I}}$, and let $\mu^\dagger$ denote the cylinder measure defined by the Gaussian field $\xi - \mathbb{E}[\xi | \Phi \xi]$, or the corresponding Gaussian measure in finite dimensions.

We say that a tuple (μ', ψ') is a saddle point of the loss function $\mathcal{E} : \mathcal{P}_2(\mathcal{B}) \times L(V, \mathcal{B}) \to \mathbb{R}$ if

$$\mathcal{E}(\mu, \psi') \leqslant \mathcal{E}(\mu', \psi') \leqslant \mathcal{E}(\mu', \psi), \quad \mu \in \mathcal{P}_2(\mathcal{B}), \ \psi \in L(V, \mathcal{B}).$$

---

[1]The *cylinder sets* of $\mathcal{B}$ consist of all sets of the form $F^{-1}(B)$ where $B \in \mathbb{R}^n$ is a Borel set and $F : \mathcal{B} \to \mathbb{R}^n$ is a continuous linear map, over all integers n. A *cylinder measure* μ, see also [82, Chap. 17], on B, is a collection of measures μ_F indexed by $F : \mathcal{B} \to \mathbb{R}^n$ over all n such that each μ_F is a Borel measure on $\mathbb{R}^n$ and such that for $F_1 : \mathcal{B} \to \mathbb{R}^{n_1}$ and $F_2 : \mathcal{B} \to \mathbb{R}^{n_2}$ and $G : \mathbb{R}^{n_1} \to \mathbb{R}^{n_2}$ linear and continuous with $F_2 = GF_1$, we have $G_* \mu_{F_1} = \mu_{F_2}$, where G_* is the pushforward operator on measures corresponding to the map G, defined by $(G_* \nu)(B) := \nu(G^{-1}B)$. When each measure μ_F is Gaussian, the cylinder measure is said to be a *Gaussian cylinder measure*. A sequence μ_n of cylinder measures such that the sequence $(\mu_n)_F$ converges in the weak topology for each F is said to converge in the *weak cylinder measure topology*.

Theorem 3.2.1 shows that the optimal strategy of Player I is the Gaussian field $\xi - \mathbb{E}[\xi|\Phi\xi]$, the optimal strategy of Player II is the conditional expectation

$$\Psi(v) = \mathbb{E}[\xi|\Phi\xi = v], \tag{3.2.2}$$

and (3.2.2) is equal to (3.1.14).

Theorem 3.2.1. *Let $\mathcal{E}$ be defined as in (3.2.1). It holds true that*

$$\max_{\mu\in\mathcal{P}_2(\mathcal{B})} \min_{\psi\in L(V,\mathcal{B})} \mathcal{E}(\mu,\psi) = \min_{\psi\in L(V,\mathcal{B})} \max_{\mu\in\mathcal{P}_2(\mathcal{B})} \mathcal{E}(\mu,\psi). \tag{3.2.3}$$

Furthermore,

- *If $\dim(V) < \infty$, then $(\mu^\dagger, \Psi)$ is a saddle point for the loss (3.2.1), where Ψ is as in (3.1.14) and (3.2.2).*

- *If $\dim(V) = \infty$, then the loss (3.2.1) admits a sequence of saddle points $(\mu_n, \Psi) \in \mathcal{P}_2(\mathcal{B}) \times L(V,\mathcal{B})$ where Ψ is as in (3.1.14) and (3.2.2), and the μ_n are Gaussian measures, with finite dimensional support, converging towards $\mu^\dagger$ in the weak cylinder measure topology.*

Proof. The proof is essentially that of [82, Thm. 18.2]   □

3.3   Gaussian Process Regression Setting

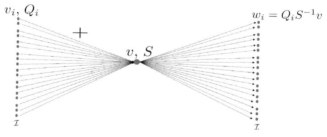

Figure 3.1: The minmax solution of the mode decomposition problem.

Let us demonstrate that Theorem 3.2.1 implies that the minmax optimal solution to Problem 3 with loss measured as the relative error in the norm (3.1.5) can be obtained via Gaussian process regression. To that end, let $\xi_i \sim \mathcal{N}(0, Q_i)$, $i \in \mathcal{I}$, be independent V_i-valued Gaussian fields defined by the norms $\|\cdot\|_{V_i}$. Recall that Q_i is defined in (3.1.6) and that ξ_i is an isometry from $(V_i^*, \|\cdot\|_{V_i^*})$ onto a Gaussian space, mapping $\phi \in V_i^*$ to $[\phi, \xi_i] \sim \mathcal{N}(0, [\phi, Q_i\phi])$. Theorem 3.2.1 asserts that the minmax estimator is (3.2.2), which, written componentwise, determines the optimal reconstruction of each mode v_j of $v = \sum_{i\in\mathcal{I}} v_i$ to be

$$\mathbb{E}\big[\xi_j\big|\sum_{i\in\mathcal{I}}\xi_i = v\big] = Q_j(\sum_{i\in\mathcal{I}}Q_i)^{-1}v, \tag{3.3.1}$$

where the right-hand side of (3.3.1) is obtained from (3.1.14), and $\sum_{i \in \mathcal{I}} Q_i$ is a shorthand notation for $\sum_i e_i Q_i e_i^*$ obtained by dropping the indications of the injections e_i and their adjoint projections e_i^*. From now on, we will use such simplified notations whenever there is no risk of confusion. In summary, the minimax solution of the abstract mode decomposition problem, illustrated in Fig. 3.1, is obtained based on the specification of the operators $Q_i : V_i^* \to V_i$ and the injections $e_i : V_i \to V$, of which the former can be interpreted as quadratic norm defining operators or as covariance operators. Table 3.1 illustrates the three equivalent interpretations—optimal recovery/operator kernel/Gaussian process regression of our methodology.

Norm	Operator/kernel	GP
$\|v_i\|_{V_i}^2 := \langle Q_i^{-1} v_i, v_i \rangle$	$Q_i : V_i^* \to V_i$	$\xi_i \sim \mathcal{N}(0, Q_i)$
$\arg\min \begin{cases} \text{minimize } \sum_i \|w_i\|_{V_i}^2 \\ \sum_i w_i = v \end{cases}$	$Q_i \left(\sum_j Q_j \right)^{-1} v$	$\mathbb{E}[\xi_i \mid \sum_j \xi_j = v]$

Table 3.1: Three equivalent interpretations—optimal recovery/operator kernel/Gaussian process regression of our methodology

Example 3.3.1. *Consider the problem of recovering the modes v_1, v_2, v_3, v_4 from the observation of the signal $v = v_1 + v_2 + v_3 + v_4$ illustrated in Fig. 3.2. In this example all modes are defined on the interval $[0,1]$, $v_1(t) = (1 + 2t^2) \cos(\theta_1(t)) - 0.5t \sin(\theta_1(t))$, $v_2(t) = 2(1-t^3) \cos(\theta_2(t)) + (-t+0.5t^2) \sin(\theta_2(t))$, $v_3(t) = 2 + t - 0.2t^2$, and v_4 is white noise (the instantiation of a centered GP with covariance function $\delta(s - t)$). $\theta_1(t) = \int_0^t \omega_1(s) \, ds$ and $\theta_2(t) = \int_0^t \omega_2(s) \, ds$ are defined by the instantaneous frequencies $\omega_1(t) = 16\pi(1 + t)$ and $\omega_2(t) = 30\pi(1 + t^2/2)$. In this recovery problem $\omega_1(t)$ and $\omega_2(t)$ are known, v_3 and the amplitudes of the oscillations of v_1 and v_2 are unknown smooth functions of time, only the distribution of v_4 is known. To define optimal recovery solutions one can either define the normed subspaces $(V_i, \|\cdot\|_{V_i})$ or (equivalently via (3.1.6)) the covariance functions/operators of the Gaussian processes ξ_i. In this example it is simpler to use the latter. To define the covariance function of the GP ξ_1 we assume that $\xi_1(t) = \zeta_{1,c}(t) \cos(\theta_1(t)) + \zeta_{1,s}(t) \sin(\theta_1(t))$, where $\zeta_{1,c}$ and $\zeta_{1,s}$ are independent identically distributed centered Gaussian processes with covariance function $\mathbb{E}[\zeta_{1,c}(s)\zeta_{1,c}(t)] = \mathbb{E}[\zeta_{1,s}(s)\zeta_{1,s}(t)] = e^{-\frac{(s-t)^2}{\gamma^2}}$ (chosen with $\gamma = 0.2$ as a prior regularity assumption). Under this choice ξ_1 is a centered GP with covariance function $K_1(s,t) = e^{-\frac{(s-t)^2}{\gamma^2}} \big(\cos(\theta_1(s)) \cos(\theta_1(t)) + \sin(\theta_1(s)) \sin(\theta_1(t)) \big)$. Note that the cosine and sine summation formulas imply that translating θ_1 by an arbitrary phase, b leaves K_1 invariant (knowing θ_1 up to a phase shift is sufficient to construct that kernel). Similarly we select the covariance function of the independent centered GP ξ_2 to be*

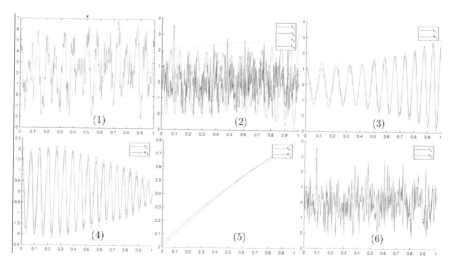

Figure 3.2: (1) The signal $v = v_1 + v_2 + v_3 + v_4$, (2) the modes v_1, v_2, v_3, v_4, (3) v_1 and its approximation w_1, (4) v_2 and its approximation w_2, (5) v_3 and its approximation w_3, (6) v_4 and its approximation w_4.

$K_2(s,t) = e^{-\frac{(s-t)^2}{\gamma^2}} \big(\cos(\theta_2(s)) \cos(\theta_2(t)) + \sin(\theta_2(s)) \sin(\theta_2(t)) \big)$. *To enforce the regularity of* ξ_3 *we select its covariance function to be* $K_3(s,t) = 1 + st + e^{-\frac{(s-t)^2}{4}}$. *Finally since* v_4 *is white noise we represent it with a centered GP with covariance function* $K_4(s,t) = \delta(s-t)$. *Figure 3.2 shows the recovered modes using* (3.3.1) *(or equivalently defined as* (3.1.14) *and the minimizer of* (3.1.4)*). In this numerical implementation the interval* $[0,1]$ *is discretized with* 302 *points (with uniform time steps between points),* ξ_4 *is a discretized centered Gaussian vector of dimension* 302 *and of identity covariance matrix and* ξ_1, ξ_2, ξ_3 *are discretized as centered Gaussian vectors with covariance matrices corresponding to the kernel matrices* $\big(K(t_i, t_j)\big)_{i,j=1}^{302}$ *corresponding to* K_1, K_2, *and* K_3 *determined by the sample points* $t_i, i = 1, \ldots, 302$.

Table 3.2 provides a summary of the approach of Example 3.3.1, illustrating the connection between the assumed mode structure and corresponding Gaussian process structure and its corresponding reproducing kernel structure.

On Additive Models   The recovery approach of Example 3.3.1 is based on the design of an appropriate additive regression model. Additive regression models are not new. They were introduced in [109] for approximating multivariate functions with sums of univariate functions. Generalized additive models (GAMs) [46] replace a linear regression model $\sum_i \alpha_i X_i$ with an additive regression model $\sum_i f_i(X_i)$ where the f_i are unspecified (smooth) functions estimated from the data. Since their inception, GAMs have become increasingly popular

Mode	GP	Kernel
$v_1(t) = a_1(t)\cos(\theta_1(t))$ θ_1 known a_1 unknown smooth	$\xi_1(t) = \zeta_1(t)\cos(\theta_1(t))$ $\mathbb{E}[\zeta_1(s)\zeta_1(t)] = e^{-\frac{|s-t|^2}{\gamma^2}}$	$K_1(s,t) = e^{-\frac{|s-t|^2}{\gamma^2}}\cos(\theta_1(s))\cos(\theta_1(t))$
$v_2(t) = a_2(t)\cos(\theta_2(t))$ θ_2 known a_2 unknown smooth	$\xi_2(t) = \zeta_2(t)\cos(\theta_2(t))$ $\mathbb{E}[\zeta_2(s)\zeta_2(t)] = e^{-\frac{|s-t|^2}{\gamma^2}}$	$K_2(s,t) = e^{-\frac{|s-t|^2}{\gamma^2}}\cos(\theta_2(s))\cos(\theta_2(t))$
v_3 unknown smooth	$\mathbb{E}[\xi_3(s)\xi_3(t)] = e^{-\frac{|s-t|^2}{\gamma^2}}$	$K_3(s,t) = e^{-\frac{|s-t|^2}{\gamma^2}}$
v_4 unknown white noise	$\mathbb{E}[\xi_4(s)\xi_4(t)] = \sigma^2\delta(s-t)$	$K_4(s,t) = \sigma^2\delta(s-t)$
$v = v_1 + v_2 + v_3 + v_4$	$\xi = \xi_1 + \xi_2 + \xi_3 + \xi_4$	$K = K_1 + K_2 + K_3 + K_4$

Table 3.2: A summary of the approach of Example 3.3.1, illustrating the connection between the assumed mode structure and corresponding Gaussian process structure and its corresponding reproducing kernel structure. Note that, for clarity of presentation, this summary does not exactly match that of Example 3.3.1

because they are both easy to interpret and easy to fit [87]. This popularity has motivated the introduction of additive Gaussian processes [27, 30] defined as Gaussian processes whose high dimensional covariance kernels are obtained from sums of low dimensional ones. Such kernels are expected to overcome the curse of dimensionality by exploiting additive non-local effects when such effects are present [30]. See Sect. 2.1. Of course, performing regression or mode decomposition with Gaussian processes (GPs) obtained as sums of independent GPs (i.e., performing kriging with kernels obtained as sums of simpler kernels) is much older since Tikhonov regularization (for signal/noise separation) has a natural interpretation as a conditional expectation $\mathbb{E}[\xi_s | \xi_s + \xi_\sigma]$ where ξ_s is a GP with a smooth prior (for the signal) and ξ_σ is a white noise GP independent from ξ_s. More recent applications include classification [72], source separation [69, 85], and the detection of the periodic part of a function from partial point evaluations [1, 29]. For that latter application, the approach of [29] is to (1) consider the RKHS H_K defined by a Matérn kernel K, (2) interpolate the data with the kernel K, and (3) recover the periodic part by projecting the interpolator (using a projection that is orthogonal with respect to the RKHS scalar product onto $H_p := \text{span}\{\cos(2\pi kt/\lambda), \sin(2\pi kt/\lambda) \mid 1 \leqslant k \leqslant q\}$ (the parameters of the Matérn kernel and the period λ are obtained via maximum likelihood estimation). Defining K_p and K_{np} as the kernels induced on H_p and its orthogonal complement in H_K, we have $K = K_p + K_{np}$ and the recovery (after MLE estimation of the parameters) can also be identified as the conditional expectation of the GP induced by K_p conditioned on the GP induced by $K_p + K_{np}$.

Chapter 4

Kernel Mode Decomposition Networks (KMDNets)

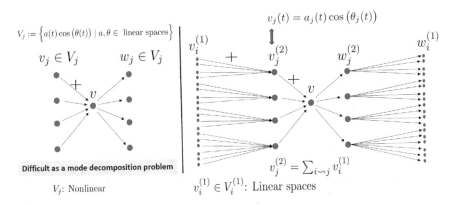

Figure 4.1: Left: Problem 1 is hard as a mode decomposition problem because the modes $v_j = a_j(t)\cos(\theta_j(t))$ live in nonlinear functional spaces. Right: one fundamental idea is to recover those modes as aggregates of finer modes v_i living in linear spaces.

In this chapter, we describe kernel mode decomposition networks (KMD-Nets) as a powerful development of the previous chapter. Indeed, the recovery approach described in Example 3.3.1 is based on the prior knowledge of (1) the number of quasi-periodic modes, (2) their phase functions θ_i, and (3) their base periodic waveform (which need not be a cosine function). In most applications, (1) and (2) are not available, and the base waveform may not be trigonometric

H. Owhadi et al., *Kernel Mode Decomposition and the Programming of Kernels*, Surveys and Tutorials in the Applied Mathematical Sciences 8, https://doi.org/10.1007/978-3-030-82171-5_4

and may not be known. Even when the base waveforms are known and trigonometric (as in Problem 1), and when the modes' phase functions are unknown, the recovery of the modes is still significantly harder than when they are known because, as illustrated in Fig. 4.1, the functional spaces defined by the modes $a_j(t) \cos\big(\theta_j(t)\big)$ (under regularity assumptions on the a_j and θ_j) are no longer linear spaces and the simple calculus of Chap. 3 requires the spaces V_j to be linear.

To address the full Problem 1, one fundamental idea is to recover those modes v_j as aggregates of finer modes v_i living in linear spaces V_i (see Fig. 4.1). In particular, we will identify i with time–frequency–phase triples (τ, ω, θ) and the spaces V_i with one dimensional spaces spanned by functions that are maximally localized in the time–frequency–phase domain (i.e., by Gabor wavelets as suggested by the approximation (1.1.1)) and recover the modes $a_j(t) \cos\big(\theta_j(t)\big)$ by aggregating the finer recovered modes.

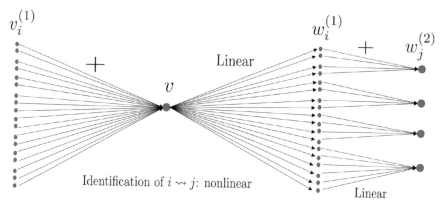

Figure 4.2: Mode decomposition/recomposition problem. Note that the nonlinearity of this model is fully represented in the identification of the relation $i \rightsquigarrow j$; once this identification is determined, all other operations are linear.

The implementation of this idea will therefore transform the nonlinear mode decomposition problem illustrated on the left-hand side of Fig. 4.1 into the mode decomposition/recomposition problem illustrated in Fig. 4.2 and transfer its nonlinearity to the identification of ancestor/descendant relationships $i \rightsquigarrow j$.

To identify these ancestor/descendant relations, we will compute the energy $E(i) := \|w_i\|_{V_i}^2$ for each recovered mode w_i, which, as illustrated in Fig. 4.3 and discussed in Sect. 4.1, can also be identified as the alignment $\langle w_i, v \rangle_{S-1}$ between recovered mode w_i and the signal v or as the alignment $\mathbb{E}[\text{Var}[\langle \xi_i, v \rangle_{S-1}]$ between the model ξ_i and the data v. Furthermore, E satisfies an energy preservation identity $\sum_i E(i) = \|v\|_{S-1}^2$, which leads to its variance decomposition interpretation. Although alignment calculations are linear, the calculations of the resulting child–ancestor relations may involve a nonlinearity (such as thresholding, graph-cut, and computation of a maximizer), and the resulting network can be seen as a sequence of sandwiched linear operations and simple nonlinear steps having striking similarities with artificial neural networks.

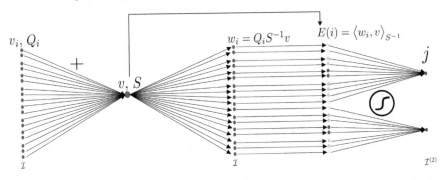

Figure 4.3: Derivation of ancestor/descendant relations from energy calculations.

Of course, this strategy can be repeated across levels of abstractions, and its complete deployment will also require the generalization of the setting of Chap. 3 (illustrated in Fig. 3.1) to a hierarchical setting (illustrated in Fig. 4.6 and described in Sect. 4.3).

4.1 Model/Data Alignment and Energy/Variance Decomposition

Using the setting and notations of Chap. 3 and fixing the observed data $v \in V$, let $E : \mathcal{I} \to \mathbb{R}_+$ be the function defined by

$$E(i) := \|\Psi_i(v)\|_{V_i}^2, \quad i \in \mathcal{I}, \tag{4.1.1}$$

where Ψ_i are the components of the optimal recovery map Ψ evaluated in Theorem 3.1.3. We will refer to $E(i)$ as the energy of the mode i in reference to its numerical analysis interpretation (motivated by the "energy" representation of $E(i) = [Q_i^{-1}\Psi_i(v), \Psi_i(v)]$ determined by (3.1.6), and the interpretation of Q_i^{-1} as an elliptic operator), and our general approach will be based on using its local and/or global maximizers to decompose/recompose kernels.

Writing $E_{\text{tot}} := \|v\|_{S^{-1}}^2$, note that (3.1.15) implies that

$$E_{\text{tot}} = \sum_{i \in \mathcal{I}} E(i). \tag{4.1.2}$$

Let $\langle \cdot, \cdot \rangle_{S^{-1}}$ be the scalar product on V defined by the norm $\| \cdot \|_{S^{-1}}$.

Proposition 4.1.1. *Let* $\xi \sim \mathcal{N}(0, Q)$ *and* $\phi := S^{-1}v$. *It holds true that for* $i \in \mathcal{I}$,

$$E(i) = \left\langle \Psi_i(v), v \right\rangle_{S^{-1}} = \text{Var}\left([\phi, \xi_i]\right) = \text{Var}\left(\langle \xi_i, v \rangle_{S^{-1}}\right). \tag{4.1.3}$$

Observe that $E(i) = \text{Var}\left(\langle \xi_i, v \rangle_{S^{-1}}\right)$ implies that $E(i)$ is a measure of the alignment between the Gaussian process (GP) model ξ_i and the data v in V and (4.1.2) corresponds to the variance decomposition

$$\text{Var}\left(\langle \sum_{i\in\mathcal{I}} \xi_i, v \rangle_{S^{-1}}\right) = \sum_{i\in\mathcal{I}} \text{Var}\left(\langle \xi_i, v \rangle_{S^{-1}}\right). \tag{4.1.4}$$

Therefore, the stronger this alignment $E(i)$ is, the better the model ξ_i is at explaining/representing the data. Consequently, we refer to the energy $E(i)$ as the *alignment energy*. Observe also that the identity $E(i) = \langle w_i, v \rangle_{S^{-1}}$ with $w_i = \Psi_i(v)$ implies that $E(i)$ is also a measure of the alignment between the optimal approximation w_i of v_i and the signal v. Table 4.1 illustrates the relations between the conservation of alignment energies and the variance decomposition derived from Theorem 3.1.3 and Proposition 4.1.1.

	Norm		Operator/kernel	GP
$E(i)$	$\|\Psi_i(v)\|_{V_i}^2 = \langle \Psi_i(v), v \rangle_{S^{-1}}$		$[S^{-1}v, Q_i S^{-1}v]$	$\text{Var}\left(\langle \xi_i, v \rangle_{S^{-1}}\right)$
$\sum_i E(i)$	$\|v\|_{S^{-1}}^2$		$[S^{-1}v, v]$	$\text{Var}\left(\langle \sum_i \xi_i, v \rangle_{S^{-1}}\right)$

Table 4.1: Identities for $E(i)$ and $\sum_i E(i)$.

4.2 Programming Modules and Feedforward Network

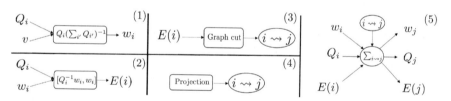

Figure 4.4: Elementary programming modules for kernel mode decomposition.

We will now combine the alignment energies of Sect. 4.1 with the mode decomposition approach of Chap. 3 to design elementary programming modules (illustrated in Fig. 4.4) for kernel mode decomposition networks (KMDNets). These will be introduced in this section and developed in the following ones.

Optimal Recovery of the Modes $(v_i)_{i\in\mathcal{I}}$ Per Chap. 3 and Theorem 3.1.3, the optimal recoveries of the modes $(v_i)_{i\in\mathcal{I}}$ given the covariance operators $(Q_i)_{i\in\mathcal{I}}$ and the observation of $\sum_{i\in\mathcal{I}} v_i$ are the elements $Q_i(\sum_{i'} Q_{i'})^{-1}v$ in V_i. This operation is illustrated in module (1) of Fig. 4.4.

Energy Function An important quantity derived from this recovery is the energy function $E : \mathcal{I} \to \mathbb{R}_+$, defined in (4.1.1) by $E(i) := [Q_i^{-1}w_i, w_i]$ with $w_i := \Psi_i(v)$ and illustrated in module (2).

Energy/Variance Decomposition Since, per (4.1.2), $E_{tot} = \sum_{i \in \mathcal{I}} E(i)$, where $E_{tot} := \|v\|_{S^{-1}}^2$ is the total energy (4.1.1), the function E can be interpreted as performing a decomposition of the total energy over the set of labels $\mathcal{I}$. When $\mathcal{I}$ can be identified with the set of vertices of a graph, the values of the $E(i)$ can be used to cut that graph into subgraphs indexed by labels $j \in \mathcal{J}$ and define a relation $i \rightsquigarrow j$ mapping $i \in \mathcal{I}$ to its subgraph j. This graph-cut operation is illustrated in module (3). Since, per Sect. 4.1, $E(i)$ is also the mean-squared alignment between the model ξ_i and the data v, and (4.1.4) is a variance decomposition, this clustering operation combines variance/model alignment information (as done with PCA) with the geometric information (as done with mixture models [75]) provided by the graph to assign a class $j \in \mathcal{J}$ to each element $i \in \mathcal{I}$.

Truncation/Projection Map The relation $i \rightsquigarrow j$ may also be obtained through a projection step, possibly ignoring the values of $E(i)$, as illustrated in module (4) (e.g., when i is an r-tuple $(i_1, i_2, \ldots, i_r)$, then the truncation/projection map $(i_1, \ldots, i_r) \rightsquigarrow (i_1, \ldots, i_{r-1})$ naturally defines a relation $\rightsquigarrow$).

Aggregation As illustrated in module (5), combining the relation $\rightsquigarrow$ with a sum $\sum_{i \rightsquigarrow j}$ produces aggregated covariance operators $Q_j := \sum_{i \rightsquigarrow j} Q_i$, modes $w_j := \sum_{i \rightsquigarrow j} w_i$, and energies $E(j) := \sum_{i \rightsquigarrow j} E(i)$ such that for $V_j := \sum_{i \rightsquigarrow j} V_i$, the modes $(w_i)_{i \rightsquigarrow j}$ are (which can be proven directly or as an elementary application of Theorem 4.3.3 in the next section) to be optimal recovery modes in $\prod_{i \rightsquigarrow j} V_i$ given the covariance operators $(Q_i)_{i \rightsquigarrow j}$ and the observation of $w_j = \sum_{i \rightsquigarrow j} w_i$ in V_j. Furthermore, we have $E(j) = [Q_j^{-1}w_j, w_j]$. Naturally, combining these elementary modules leads to more complex secondary modules (illustrated in Fig. 4.5) whose nesting produces a network aggregating the fine modes w_i into increasingly coarse modes with the last node corresponding to v.

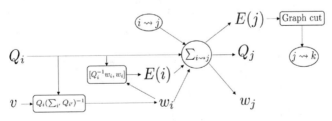

Figure 4.5: Programming modules derived from the elementary modules of Fig. 4.4.

4.3 Hierarchical Mode Decomposition

We now describe how a hierarchy of mode decomposition/recomposition steps discussed in Sect. 4.2 naturally produces a hierarchy of labels, covariance operators, subspaces, and recoveries (illustrated in Fig. 4.6) along with important geometries and inter-relationships. This description will lead to the meta-algorithm, Algorithm 1, presented in Sect. 4.4, aimed at the production of a KMDNet such as the one illustrated in Fig. 4.6. Section 4.5 will present a practical application to Problem 1.

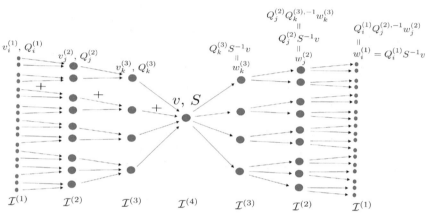

Figure 4.6: The generalization of abstract mode decomposition problem of Fig. 3.1 to a hierarchy as described in Sect. 4.3.

Our first step is to generalize the recovery approach of Chap. 3 to the case where V is the sum of a hierarchy of linear nested subspaces labeled by a hierarchy of indices, as defined below.

Definition 4.3.1. *For $q \in \mathbb{N}^*$, let $\mathcal{I}^{(1)}, \ldots, \mathcal{I}^{(q)}$ be finite sets of indices such that $\mathcal{I}^{(q)} = \{1\}$ has only one element. Let $\cup_{l=1}^q \mathcal{I}^{(l)}$ be endowed with a relation $\rightsquigarrow$ that is (1) transitive, i.e., $i \rightsquigarrow j$ and $j \rightsquigarrow k$ implies $i \rightsquigarrow k$, (2) directed, i.e., $i \in \mathcal{I}^{(s)}$ and $j \in \mathcal{I}^{(r)}$ with $s \geqslant r$ implies $i \not\rightsquigarrow j$ (that is, i does not lead to j), and (3) locally surjective, i.e., any element $j \in \mathcal{I}^{(r)}$ with $r > 1$ has at least one $i \in \mathcal{I}^{(r-1)}$ such that $i \rightsquigarrow j$. For $1 \leqslant k < r \leqslant q$ and an element $i \in \mathcal{I}^{(r)}$, write $i^{(k)} := \{j \in \mathcal{I}^{(k)} \mid j \rightsquigarrow i\}$ for the level k ancestors of i.*

Let $V_i^{(k)}$, $i \in \mathcal{I}^{(k)}, k \in \{1, \ldots, q\}$, be a hierarchy of nested linear subspaces of a separable Hilbert space V such that

$$V_1^{(q)} = V$$

and, for each level in the hierarchy $k \in \{1, \ldots, q-1\}$,

$$V_i^{(k+1)} = \sum_{j \in i^{(k)}} V_j^{(k)}, \qquad i \in \mathcal{I}^{(k+1)}. \tag{4.3.1}$$

Let $\mathcal{B}^{(q)} = V$, and for $k \in \{1, \ldots, q-1\}$, let $\mathcal{B}^{(k)}$ be the product space

$$\mathcal{B}^{(k)} := \prod_{i \in \mathcal{I}^{(k)}} V_i^{(k)} . \tag{4.3.2}$$

For $k < r$ and $j \in \mathcal{I}^{(r)}$, let

$$\mathcal{B}_j^{(k)} := \prod_{i \in j^{(k)}} V_i^{(k)} , \tag{4.3.3}$$

and let

$$\Phi_j^{(r,k)} : \mathcal{B}_j^{(k)} \to V_j^{(r)}$$

be defined by

$$\Phi_j^{(r,k)}(u) := \sum_{i \in j^{(k)}} u_i, \qquad u \in \mathcal{B}_j^{(k)} . \tag{4.3.4}$$

Putting these components together as $\Phi^{(r,k)} = (\Phi_j^{(r,k)})_{j \in \mathcal{I}^{(r)}}$, we obtain the multi-linear map

$$\Phi^{(r,k)} : \mathcal{B}^{(k)} \to \mathcal{B}^{(r)}, \quad 1 \leqslant k < r \leqslant q,$$

defined by

$$\Phi^{(r,k)}(u) := \Big(\sum_{i \in j^{(k)}} u_i \Big)_{j \in \mathcal{I}^{(r)}}, \qquad u = (u_i)_{i \in \mathcal{I}^{(k)}} \in \mathcal{B}^{(k)} . \tag{4.3.5}$$

To put hierarchical metric structure on these spaces, for $k \in \{1, \ldots, q\}$ and $i \in \mathcal{I}^{(k)}$, let

$$Q_i^{(k)} : V_i^{(k),*} \to V_i^{(k)}$$

be positive symmetric linear bijections determining the quadratic norms

$$\|v\|_{V_i^{(k)}}^2 = [Q_i^{(k),-1} v, v], \qquad v \in V_i^{(k)}, \tag{4.3.6}$$

on the $V_i^{(k)}$. Then for $k \in \{1, \ldots, q\}$, let $\mathcal{B}^{(k)}$ be endowed with the quadratic norm defined by

$$\|u\|_{\mathcal{B}^{(k)}}^2 = \sum_{i \in \mathcal{I}^{(k)}} \|u_i\|_{V_i^{(k)}}^2, \qquad u \in \mathcal{B}^{(k)} , \tag{4.3.7}$$

and, for $k < r \leqslant q$ and $j \in \mathcal{I}^{(r)}$, let $\mathcal{B}_j^{(k)} := \prod_{i \in j^{(k)}} V_i^{(k)}$ be endowed with the quadratic norm defined by

$$\|u\|_{\mathcal{B}_j^{(k)}}^2 = \sum_{i \in j^{(k)}} \|u_i\|_{V_i^{(k)}}^2, \qquad u \in \mathcal{B}_j^{(k)}.$$

For $1 \leqslant k < r \leqslant q$, the nesting relations (4.3.1) imply that

$$V_i^{(k)} \subset V_j^{(r)}, \quad i \in j^{(k)}, \ j \in \mathcal{I}^{(r)},$$

so that the subset injection

$$e_{j,i}^{(r,k)} : V_i^{(k)} \to V_j^{(r)} \tag{4.3.8}$$

is well defined for all $i \in j^{(k)}$, $j \in \mathcal{I}^{(r)}$, and since all spaces are complete, they have well-defined adjoints, which we write

$$e_{i,j}^{(k,r)} : V_j^{(r),*} \to V_i^{(k),*}. \tag{4.3.9}$$

For $1 \leqslant k < r \leqslant q$, $i \in \mathcal{I}^{(k)}$ and $j \in \mathcal{I}^{(r)}$, let

$$\Psi_{i,j}^{(k,r)} : V_j^{(r)} \to V_i^{(k)}$$

be defined by

$$\Psi_{i,j}^{(k,r)}(v_j) = Q_i^{(k)} e_{i,j}^{(k,r)} Q_j^{(r),-1} v_j, \qquad v_j \in V_j^{(r)}, \tag{4.3.10}$$

so that, when putting the components together as

$$\Psi_j^{(k,r)} := (\Psi_{i,j}^{(k,r)})_{i \in j^{(k)}}, \tag{4.3.11}$$

(4.3.3) determines the multi-linear map

$$\Psi_j^{(k,r)} : V_j^{(r)} \to \mathcal{B}_j^{(k)}.$$

Further collecting components simultaneously over the range and domain as

$$\Psi^{(k,r)} = (\Psi_j^{(k,r)})_{j \in \mathcal{I}^{(r)}},$$

we obtain from (4.3.2) the multi-linear map

$$\Psi^{(k,r)} : \mathcal{B}^{(r)} \to \prod_{j \in \mathcal{I}^{(r)}} \mathcal{B}_j^{(k)}$$

defined by

$$\Psi^{(k,r)}(v) = \left(Q_i^{(k)} e_{i,j}^{(k,r)} Q_j^{(r),-1} v_j \right)_{i \in j^{(k)}}, \qquad v = (v_j)_{j \in \mathcal{I}^{(r)}} \in \mathcal{B}^{(r)}. \tag{4.3.12}$$

The following condition assumes that the relation $\leadsto$ determines a mapping $\leadsto : \mathcal{I}^{(k)} \to \mathcal{I}^{(k+1)}$ for all $k = 1, \ldots, q-1$.

Condition 4.3.2. *For $k \in \{1, \ldots, q-1\}$, every $i \in \mathcal{I}^{(k)}$ has a unique descendant in $\mathcal{I}^{(k+1)}$. That is, there exists a $j \in \mathcal{I}^{(k+1)}$ with $i \leadsto j$ and there is no other $j' \in \mathcal{I}^{(k+1)}$ such that $i \leadsto j'$.*

Condition 4.3.2 simplifies the previous results as follows: the subsets $(\{i \in j^{(k)}\})_{j \in \mathcal{I}^{(k+1)}}$ form a partition of $\mathcal{I}^{(k)}$, so that, for $k < r$, we obtain the simultaneous product structure

$$\mathcal{B}^{(k)} = \prod_{j \in \mathcal{I}^{(r)}} \mathcal{B}_j^{(k)}$$

$$\mathcal{B}^{(r)} = \prod_{j \in \mathcal{I}^{(r)}} V_i^{(r)}, \tag{4.3.13}$$

so that both

$$\Phi^{(k,r)} : \mathcal{B}^{(k)} \to \mathcal{B}^{(r)}$$

and

$$\Psi^{(k,r)} : \mathcal{B}^{(r)} \to \mathcal{B}^{(k)}$$

are diagonal multi-linear maps with components

$$\Phi_j^{(r,k)} : \mathcal{B}_j^{(k)} \to V_j^{(r)}$$

and

$$\Psi_j^{(k,r)} : V_j^{(r)} \to \mathcal{B}_j^{(k)},$$

respectively. Moreover, both maps are *linear* under the isomorphism between products and external direct sums of vector spaces. For $r > k$, we have the following connections between $\mathcal{B}^{(k)}, \mathcal{B}^{(r)}, V_i^{(k)}$, and $V_j^{(r)}$:

$$
\begin{array}{ccc}
\mathcal{B}^{(k)} & \xleftarrow{\quad \prod_{i \in \mathcal{I}^{(k)}} \quad} & V_i^{(k)} \\[2mm]
\Psi^{(k,r)} \big\uparrow \quad \big\downarrow \Phi^{(r,k)} & & \big\downarrow \Sigma_{i \in j^{(k)}} \\[2mm]
\mathcal{B}^{(r)} & \xleftarrow{\quad \prod_{j \in \mathcal{I}^{(r)}} \quad} & V_j^{(r)}
\end{array}
\tag{4.3.14}
$$

The following theorem is a consequence of Theorem 3.1.3.

Theorem 4.3.3. *Assume that Condition 4.3.2 holds and that the* $Q_i^{(k)} : V_i^{(k),*} \to : V_i^{(k)}$ *satisfy the nesting relations*

$$Q_j^{(k+1)} = \sum_{i \in j^{(k)}} e_{j,i}^{(k+1,k)} Q_i^{(k)} e_{i,j}^{(k,k+1)}, \quad j \in \mathcal{I}^{(k+1)}, \tag{4.3.15}$$

for $k \in \{1, \ldots, q-1\}$. *Then, for* $1 \leqslant k < r \leqslant q$,

- $\Psi^{(k,r)} \circ \Phi^{(r,k)} (u)$ *is the minmax recovery of* $u \in \mathcal{B}^{(k)}$ *given the observation of* $\Phi^{(r,k)}(u) \in \mathcal{B}^{(r)}$ *using the relative error in* $\| \cdot \|_{\mathcal{B}^{(k)}}$ *norm as a loss.*

- $\Phi^{(r,k)} \circ \Psi^{(k,r)}$ *is the identity map on* $\mathcal{B}^{(r)}$.

- $\Psi^{(k,r)} : (\mathcal{B}^{(r)}, \| \cdot \|_{\mathcal{B}^{(r)}}) \to (\mathcal{B}^{(k)}, \| \cdot \|_{\mathcal{B}^{(k)}})$ *is an isometry.*

- $\Phi^{(k,r),*} : (\mathcal{B}^{(r),*}, \| \cdot \|_{\mathcal{B}^{(r),*}}) \to (\mathcal{B}^{(k),*}, \| \cdot \|_{\mathcal{B}^{(k),*}})$ *is an isometry.*

Moreover, we have the following semigroup properties for $1 \leqslant k < r < s \leqslant q$:

- $\Phi^{(s,k)} = \Phi^{(s,r)} \circ \Phi^{(r,k)}$,

- $\Psi^{(k,s)} = \Psi^{(k,r)} \circ \Psi^{(r,s)}$,

- $\Psi^{(r,s)} = \Phi^{(r,k)} \circ \Psi^{(k,s)}$.

Remark 4.3.4. *The proof of Theorem 4.3.3 also demonstrates that, under its assumptions, for $1 \leqslant k < r \leqslant q$ and $j \in \mathcal{I}^{(r)}$, $\Psi_j^{(k,r)} \circ \Phi_j^{(r,k)}(u)$ is the minmax recovery of $u \in \mathcal{B}_j^{(k)}$ given the observation of $\Phi_j^{(r,k)}(u) \in V_j^{(r)}$ using the relative error in $\| \cdot \|_{\mathcal{B}_j^{(k)}}$ norm as a loss. Furthermore, $\Phi_j^{(r,k)} \circ \Psi_j^{(k,r)}$ is the identity map on $V_j^{(r)}$ and $\Psi_j^{(k,r)} : (V_j^{(r)}, \| \cdot \|_{V_j^{(r)}}) \to (\mathcal{B}_j^{(k)}, \| \cdot \|_{\mathcal{B}_j^{(k)}})$ and $\Phi_j^{(k,r),*} : (V_j^{(r),*}, \| \cdot \|_{V_j^{(r),*}}) \to (\mathcal{B}_j^{(k),*}, \| \cdot \|_{\mathcal{B}_j^{(k),*}})$ are isometries.*

Gaussian Process Regression Interpretation   As in the setting of Sect. 4.3, for $k \in \{1, \ldots, q\}$, let

$$Q^{(k)} : \mathcal{B}^{(k),*} \to \mathcal{B}^{(k)}$$

be the block-diagonal operator

$$Q^{(k)} := \mathrm{diag}(Q_i^{(k)})_{i \in \mathcal{I}^{(k)}}$$

defined by its action $Q^{(k)}\phi := (Q_i^{(k)}\phi_i)_{i \in \mathcal{I}^{(k)}}$, $\phi \in \mathcal{B}^{(k),*}$, and, as discussed in Sect. 3.2, write

$$\xi^{(k)} \sim \mathcal{N}(0, Q^{(k)})$$

for the centered Gaussian field on $\mathcal{B}^{(k)}$ with covariance operator $Q^{(k)}$.

Theorem 4.3.5. *Under the assumptions of Theorem 4.3.3, for $1 < k \leqslant q$, the distribution of $\xi^{(k)}$ is that of $\Phi^{(k,1)}(\xi^{(1)})$. Furthermore, $\xi^{(1)}$ conditioned on $\Phi^{(k,1)}(\xi^{(1)})$ is a time reverse martingale[1] in k and, for $1 \leqslant k < r \leqslant q$, we have*

$$\Psi^{(k,r)}(v) = \mathbb{E}\big[\xi^{(k)} \mid \Phi^{(r,k)}(\xi^{(k)}) = v\big], \qquad v \in \mathcal{B}^{(r)}. \tag{4.3.16}$$

4.4   Mode Decomposition Through Partitioning and Integration

In the setting of Sect. 4.3, recall that $\mathcal{I}^{(q)} = \{1\}$ and $V_1^{(q)} = V$ so that the index j in $\Psi_{i,j}^{(k,q)}$ defined in (4.3.10) only has one value $j = 1$ and $1^{(k)} = \mathcal{I}^{(k)}$, and therefore

$$\Psi_{i,1}^{(k,q)}(v) := Q_i^{(k)} e_{i,1}^{(k,q)} Q_1^{(q),-1} v, \qquad v \in V, i \in \mathcal{I}^{(k)}. \tag{4.4.1}$$

Fix a $v \in V$, and for $k \in \{1, \ldots, q\}$, let

$$E^{(k)} : \mathcal{I}^{(k)} \to \mathbb{R},$$

defined by

$$E^{(k)}(i) := \big\| \Psi_{i,1}^{(k,q)}(v) \big\|_{V_i^{(k)}}^2, \quad i \in \mathcal{I}^{(k)}, \tag{4.4.2}$$

---

[1]If $\mathcal{F}_n$ is a decreasing sequence of sub-σ fields of a σ-field $\mathcal{F}$ and Y is an $\mathcal{F}$ measurable random variable, then $(X_n, \mathcal{F}_n)$, where $E_n := \mathbb{E}[Y|\mathcal{F}_n]$ is a reverse martingale, in that $\mathbb{E}[X_n|\mathcal{F}_{n+1}] = X_{n+1}$.

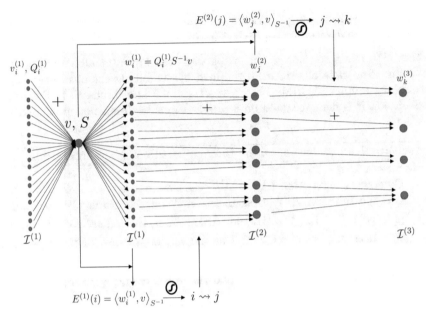

Figure 4.7: Derivation of the hierarchy from alignments.

be the alignment energy of the mode $i \in \mathcal{I}^{(k)}$. Under the nesting relations
(4.3.15), the definition (4.3.6) of the norms and the semigroup properties of
the subspace embeddings (4.3.8) imply that

$$E^{(k+1)}(i) = \sum_{i' \in i^{(k)}} E^{(k)}(i'), \quad i \in \mathcal{I}^{(k+1)}, \; k \in \{1, \ldots, q-1\}. \qquad (4.4.3)$$

We will now consider applications where the space $(V, \|\cdot\|_V)$ is known, and
the spaces $(V_i^{(1)}, \|\cdot\|_{V_i^{(1)}})$, including their index set $\mathcal{I}^{(1)}$, are known, but the
spaces $(V_j^{(k)}, \|\cdot\|_{V_j^{(k)}})$ and their indices $\mathcal{I}^{(k)}$ are unknown for $1 < k < q$, as is any
relation $\rightsquigarrow$ connecting them. Instead, they will be constructed by induction
from model/data alignments as illustrated in Figs. 4.3 and 4.7 and explained
below. In these applications

$$(V, \|\cdot\|_V) = (V_1^{(q)}, \|\cdot\|_{V_1^{(q)}}),$$

$V = \sum_{i \in \mathcal{I}^{(1)}} V_i^{(1)}$ and the operator $Q_1^{(q)} : V^* \to V$ associated with the norm
$\|\cdot\|_{V_1^{(q)}}$ is the sum

$$Q_1^{(q)} = \sum_{i \in \mathcal{I}^{(1)}} e_{1,i}^{(q,1)} Q_i^{(1)} e_{i,1}^{(1,q)}. \qquad (4.4.4)$$

In this construction, we assume that the set of indices $\mathcal{I}^{(1)}$ are vertices of
a graph $G^{(1)}$, whose edges provide neighbor relations among the indices. The

following meta-algorithm, Algorithm 1, forms a general algorithmic framework for the adaptive determination of the intermediate spaces $(V_j^{(k)}, \|\cdot\|_{V_j^{(k)}})$, their indices $\mathcal{I}^{(k)}$, and a relation $\leadsto$, in such a way that Theorem 4.3.3 applies. Observe that this meta-algorithm is obtained by combining the elementary programming modules illustrated in Figs. 4.4 and 4.5 and discussed in Sect. 4.2. In Sect. 4.5, it is demonstrated on a problem in time–frequency mode decomposition.

Algorithm 1 Mode decomposition through partitioning and integration

---
1: **for** $k = 1$ to $q - 2$ **do**
2:    Compute the function $E^{(k)} : \mathcal{I}^{(k)} \to \mathbb{R}_+$ defined by (4.4.1) and (4.4.2).
3:    Use the function $E^{(k)}$ to segment/partition the graph $G^{(k)}$ into subgraphs $(G_j^{(k+1)})_{j \in \mathcal{I}^{(k+1)}}$, thereby determining the indices $\mathcal{I}^{(k+1)}$. Define the ancestors $j^{(k)}$ of $j \in \mathcal{I}^{(k+1)}$ as the vertices $i \in \mathcal{I}^{(k)}$ of the sub-graph $G_j^{(k+1)}$.
4:    Identify the subspaces $V_j^{(k+1)}$ and the operators $Q_j^{(k+1)}$ through (4.3.1) and (4.3.15).
5: **end for**
6: Recover the modes $(\Psi_i^{(q-1,q)}(v))_{i \in \mathcal{I}^{(q-1)}}$ of v.

4.5 Application to Time–Frequency Decomposition

We will now propose a solution to Problem 1 based on the hierarchical segmentation approach described in Sect. 4.4. We will employ the GPR interpretation of Sect. 3.3 and assume that the noisy signal $v = u + v_\sigma$, where v_σ is the noise, is the realization of a Gaussian process ξ obtained by integrating Gabor wavelets [37] against white noise. To that end, for $\tau, \theta \in \mathbb{R}$ and $\omega, \alpha > 0$, let

$$\chi_{\tau,\omega,\theta}(t) := \left(\frac{2}{\pi^3}\right)^{\frac{1}{4}} \sqrt{\frac{\omega}{\alpha}} \cos\big(\omega(t - \tau) + \theta\big) e^{-\frac{\omega^2(t-\tau)^2}{\alpha^2}}, \qquad t \in \mathbb{R}, \qquad (4.5.1)$$

be the shifted/scaled Gabor wavelet, whose scaling is motivated by the normalization $\int_{-\pi}^{\pi} \int_{\mathbb{R}} \chi_{\tau,\omega,\theta}^2(t)\, dt\, d\theta = 1$; see Fig. 4.8 for an illustration of the Gabor wavelets. Recall [37] that each χ is minimally localized in the time–frequency domain (it minimizes the product of standard deviations in the time and frequency domains) and the parameter α is proportional to the ratio between localization in frequency and localization in space.

Let $\zeta(\tau, \omega, \theta)$ be a white noise process on $\mathbb{R}^3$ (a centered GP with covariance function $\mathbb{E}\big[\zeta(\tau, \omega, \theta)\zeta(\tau', \omega', \theta')\big] = \delta(\tau - \tau')\delta(\omega - \omega')\delta(\theta - \theta')$), and let

$$\xi_u(t) := \int_{-\pi}^{\pi} \int_{\omega_{\min}}^{\omega_{\max}} \int_0^1 \zeta(\tau, \omega, \theta)\chi_{\tau,\omega,\theta}(t)d\tau\, d\omega\, d\theta, \quad t \in \mathbb{R}. \qquad (4.5.2)$$

Letting, for each τ, ω, and θ,

$$K_{\tau,\omega,\theta}(s, t) := \chi_{\tau,\omega,\theta}(s)\chi_{\tau,\omega,\theta}(t), \quad s, t \in \mathbb{R}, \qquad (4.5.3)$$

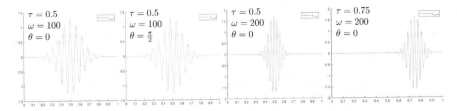

Figure 4.8: Gabor wavelets $\chi_{\tau,\omega,\theta}$ (4.5.1) for various parameter values with $\alpha = 16$.

be the reproducing kernel associated with the wavelet $\chi_{\tau,\omega,\theta}$, it follows that ξ_u is a centered GP with covariance function

$$K_u(s,t) = \int_{-\pi}^{\pi} \int_{\omega_{\min}}^{\omega_{\max}} \int_0^1 K_{\tau,\omega,\theta}(s,t) d\tau \, d\omega \, d\theta, \quad s,t \in \mathbb{R}. \qquad (4.5.4)$$

Given $\sigma > 0$, let $\xi_\sigma(t)$ be a white noise process on $\mathbb{R}$ (independent from ζ) of variance σ^2 (a centered GP with covariance function $\mathbb{E}\left[\xi_\sigma(s)\xi_\sigma(t)\right] = \sigma^2\delta(s-t)$), and let ξ, the GP defined by

$$\xi := \xi_u + \xi_\sigma, \qquad (4.5.5)$$

be used to generate the observed signal $v = u + v_\sigma$. ξ is a centered GP with covariance function defined by the kernel

$$K := K_u + K_\sigma \qquad (4.5.6)$$

with

$$K_\sigma(s,t) = \sigma^2\delta(s-t). \qquad (4.5.7)$$

Hence, compared to the setting of Chap. 3, and apart from the mode corresponding to the noise ξ_σ, the finite number of modes indexed by $\mathcal{I}$ has been turned into a continuum of modes indexed by

$$\mathcal{I} := \left\{(\tau,\omega,\theta) \in [0,1] \times [\omega_{\min},\omega_{\max}] \times (-\pi,\pi]\right\}$$

with corresponding one dimensional subspaces

$$V_{(\tau,\omega,\theta)}^{(1)} = \mathrm{span}\{\chi_{\tau,\omega,\theta}\},$$

positive operators $Q_{\tau,\omega,\theta}$ defined by the kernels $K_{\tau,\omega,\theta}(s,t)$ and the integral

$$K_u(s,t) = \int_{-\pi}^{\pi} \int_{\omega_{\min}}^{\omega_{\max}} \int_0^1 K_{\tau,\omega,\theta}(s,t) d\tau \, d\omega \, d\theta, \quad s,t \in \mathbb{R},$$

of these kernels (4.5.4) to obtain a master kernel K_u instead of a sum

$$S = \sum_{i \in \mathcal{I}} e_i Q_i e_i^*$$

as in (3.1.11). Table 4.2 illustrates the time–frequency version of Table 3.2 we have just developed, and the following remark explains the connection between kernels and operators in more detail.

Remark 4.5.1 (Kernels, Operators, and Discretizations). *This kernel mode decomposition framework constructs reproducing kernels K through the integration of elementary reproducing kernels, but the recovery formula of Theorem 3.1.3 requires the application of operators, and their inverses, corresponding to these kernels. In general, there is no canonical connection between kernels and operators, but here we consider restricting to the unit interval $[0,1] \subset \mathbb{R}$ in the time variable t. Then, each kernel K under consideration other than K_σ corresponds to the symmetric positive integral operator*

$$\bar{K} : L^2[0,1] \to L^2[0,1]$$

Mode	GP	Kernel
$v_{\tau,\omega,\theta}(t) = a_{\tau,\omega,\theta}(t)\chi_{\tau,\omega,\theta}(t)$	$\xi_{\tau,\omega,\theta}(t) = \zeta(\tau,\omega,\theta)\chi_{\tau,\omega,\theta}(t)$	$K_{\tau,\omega,\theta}(s,t) = \chi_{\tau,\omega,\theta}(s)\chi_{\tau,\omega,\theta}(t)$
$a_{\tau,\omega,\theta}$ unknown in L^2	$\mathbb{E}[\zeta(\tau,\omega,\theta)\zeta(\tau',\omega',\theta')]$ $= \delta(\tau-\tau')\delta(\omega-\omega')\delta(\theta-\theta')$	
$v_{\tau,\omega} = \int_{-\pi}^{\pi} v_{\tau,\omega,\theta}\,d\theta$	$\xi_{\tau,\omega}(t) = \int_{-\pi}^{\pi} \xi_{\tau,\omega,\theta}(t)\,d\theta$	$K_{\tau,\omega}(s,t) =$ $\int_{-\pi}^{\pi} K_{\tau,\omega,\theta}(s,t)\,d\theta$
$v_u = \int\int\int v_{\tau,\omega,\theta}\,d\tau\,d\omega\,d\theta$	$\xi_u(t) = \int\int\int \xi_{\tau,\omega,\theta}(t)\,d\tau\,d\omega\,d\theta$	$K_u(s,t)$ $= \int\int\int K_{\tau,\omega,\theta}(s,t)\,d\tau\,d\omega\,d\theta$
v_σ unknown white noise	$\mathbb{E}[\xi_\sigma(s)\xi_\sigma(t)] = \sigma^2\delta(s-t)$	$K_\sigma(s,t) = \sigma^2\delta(s-t)$
$v = v_u + v_\sigma$	$\xi = \xi_u + \xi_\sigma$	$K = K_u + K_\sigma$
$v_i = \int_{A(i)} v_{\tau,\omega}\,d\tau\,d\omega$	$\xi_i = \int_{A(i)} \xi_{\tau,\omega}\,d\tau\,d\omega$	$K_i = \int_{A(i)} K_{\tau,\omega}\,d\tau\,d\omega$

Table 4.2: The time–frequency version of Table 3.2.

defined by

$$\left(\bar{K}f\right)(s) := \int_0^1 K(s,t)f(t)\,dt, \quad s \in [0,1], \ f \in L^2[0,1].$$

Moreover, these kernels all have sufficient regularity that $\bar{K}$ is compact and therefore not invertible, see e.g., Steinwart and Christmann [107, Thm. 4.27]. On the other hand, the operator

$$\bar{K}_\sigma : L^2[0,1] \to L^2[0,1]$$

corresponding to the white noise kernel K_σ (4.5.7) is

$$\bar{K}_\sigma = \sigma^2 I,$$

where

$$I : L^2[0,1] \to L^2[0,1]$$

is the identity map. Since $K = K_u + K_\sigma$ (4.5.6), the operator $\bar{K} = \bar{K}_u + \bar{K}_\sigma$ is a symmetric positive compact operator plus a positive multiple of the identity, and therefore it is Fredholm and invertible. Consequently, we can apply Theorem 3.1.3 for the optimal recovery.

In addition, in numerical applications, τ and ω are discretized (using $N+1$ discretization steps) and the integrals in (4.5.11) are replaced by sums over $\tau_k := k/N$ and $\omega_k := \omega_{\min} + \frac{k}{N}(\omega_{\max} - \omega_{\min})$ ($k \in \{0, 1, \ldots, N\}$). Moreover, as in Example 3.3.1, the time interval $[0,1]$ is discretized into M points and the corresponding operators on $\mathbb{R}^M$ are $\sigma^2 I$, where $I : \mathbb{R}^M \to \mathbb{R}^M$ is the identity, plus the kernel matrix $(K_u(t_i, t_j))_{i,j=1}^M$ corresponding to the sample points $t_i, i = 1, \ldots, M$.

For simplicity and conciseness, henceforth, we will keep describing the proposed approach in the continuous setting. Moreover, except in Sect. 4.6, we will use overload notation and not use the $\bar{K}$ notation, but instead use the same symbol K for a kernel and its corresponding operator.

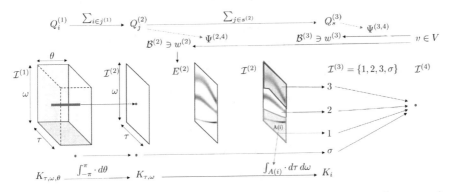

Figure 4.9: Mode decomposition through partitioning and integration. $q = 4$, $w^{(3)} := \Psi^{(3,4)}v$, $w^{(2)} := \Psi^{(2,4)}v$, and σ corresponds to the noise component.

We now describe the hierarchical approach of Sect. 4.4 to this time–frequency setting and illustrate it in Fig. 4.9. To that end, we identify $\mathcal{I}$ with $\mathcal{I}^{(1)}$ so that

$$\mathcal{I}^{(1)} = \{(\tau, \omega, \theta) \in [0,1] \times [\omega_{\min}, \omega_{\max}] \times (-\pi, \pi]\} \cup \{\sigma\},$$

where the noise mode has been illustrated in Fig. 4.9 by adding an isolated point with label σ to each set $\mathcal{I}^{(k)}$ with $k < q = 4$.

Although Line 3 of Algorithm 1 uses the energy $E^{(1)}$ at level $k = 1$ to partition the index set $\mathcal{I}^{(1)}$, the algorithm is flexible with regard to if or how we use it. In this particular application, we first ignore the computation of $E^{(1)}$ and straightforward partition $\mathcal{I}^{(1)}$ into a family of subsets

$$\mathcal{I}^{(1)}_{\tau,\omega} := \left\{(\tau,\omega,\theta) : \theta \in (-\pi,\pi]\right\} \cup \{\sigma\}, \qquad (\tau,\omega) \in [0,1] \times [\omega_{\min}, \omega_{\max}],$$

indexed by τ and ω, so that the corresponding index set at level $k = 2$ is

$$\mathcal{I}^{(2)} = \left\{(\tau,\omega) \in [0,1] \times [\omega_{\min}, \omega_{\max}]\right\} \cup \{\sigma\},$$

and the ancestors of (τ,ω,σ) are

$$(\tau,\omega,\sigma)^{(2)} = \left\{(\tau,\omega,\theta) : \theta \in (-\pi,\pi]\right\} \cup \{\sigma\}.$$

The subspace corresponding to the label (τ,ω) is then

$$V^{(2)}_{(\tau,\omega)} = \operatorname{span}\{\chi_{\tau,\omega,\theta} \mid \theta \in (-\pi,\pi]\},$$

and, as in (4.3.15), its associated positive operator is characterized by the kernel

$$K_{\tau,\omega} := \int_{-\pi}^{\pi} K_{\tau,\omega,\theta} d\theta. \tag{4.5.8}$$

We can evaluate $K_{\tau,\omega}$ using (4.5.3) and (4.5.1) by defining

$$\chi_{\tau,\omega,c}(t) \quad := \quad \left(\frac{2}{\pi}\right)^{\frac{1}{4}} \sqrt{\frac{\omega}{\alpha}} \cos(\omega(t-\tau))e^{-\frac{\omega^2(t-\tau)^2}{\alpha^2}}, \quad t \in \mathbb{R},$$

$$\chi_{\tau,\omega,s}(t) \quad := \quad \left(\frac{2}{\pi}\right)^{\frac{1}{4}} \sqrt{\frac{\omega}{\alpha}} \sin(\omega(t-\tau))e^{-\frac{\omega^2(t-\tau)^2}{\alpha^2}}, \quad t \in \mathbb{R}, \tag{4.5.9}$$

and using the cosine summation formula to obtain

$$K_{\tau,\omega}(s,t) := \chi_{\tau,\omega,c}(s)\chi_{\tau,\omega,c}(t) + \chi_{\tau,\omega,s}(s)\chi_{\tau,\omega,s}(t). \tag{4.5.10}$$

Therefore, $V^{(2)}_{(\tau,\omega)} = \operatorname{span}\{\chi_{\tau,\omega,c}, \chi_{\tau,\omega,s}\}$ and (4.5.4) reduces to

$$K_u(s,t) = \int_{\omega_{\min}}^{\omega_{\max}} \int_0^1 K_{\tau,\omega}(s,t)d\tau \, d\omega. \tag{4.5.11}$$

Using $K := K_u + K_\sigma$ (4.5.6), let f be the solution of the linear system $\int_0^1 K(s,t)f(t)\,dt = v(s)$, i.e.,

$$Kf = v, \tag{4.5.12}$$

and let $E(\tau,\omega)$ be the energy of the recovered mode indexed by (τ,ω), i.e.,

$$E(\tau,\omega) = \int_0^1 \int_0^1 f(s)K_{\tau,\omega}(s,t)f(t)\,ds\,dt, \qquad (\tau,\omega) \in [0,1] \times [\omega_{\min}, \omega_{\max}]. \tag{4.5.13}$$

Since $Kf = v$ implies that

$$v^T K^{-1} v = f^T K f,$$

it follows that

$$v^T K^{-1} v = \int_{\omega_{\min}}^{\omega_{\max}} \int_0^1 E(\tau, \omega) \, d\tau \, d\omega + f^T K_\sigma f. \tag{4.5.14}$$

For the recovery of the m (which is unknown) modes using Algorithm 1, at the second level $k = 2$ we use $E(\tau, \omega)$ to partition the time–frequency domain of (τ, ω) into n disjoint subsets $A(1), A(2), \ldots, A(n)$. As illustrated in Fig. 4.9, $n = 3$ is determined from $E(\tau, \omega)$, and $\mathcal{I}^{(3)}$ is defined as $\{1, 2, \ldots, n, \sigma\}$, the subspace corresponding to the mode $i \neq \sigma$ as $V_i^{(3)} = \text{span}\{\chi_{\tau, \omega, c}, \chi_{\tau, \omega, s} \mid (\tau, \omega) \in A(i)\}$ and the kernel associated with the mode $i \neq \sigma$ as

$$K_i(s, t) = \int_{(\tau, \omega) \in A(i)} K_{\tau, \omega}(s, t) d\tau \, d\omega, \qquad s, t \in \mathbb{R}, \tag{4.5.15}$$

as displayed in the bottom row in Table 4.2, so that

$$K_u = \sum_{i=1}^n K_i.$$

We then apply the optimal recovery formula of Theorem 3.1.3 to approximate the modes of $v_1, \ldots, v_n$ of u from the noisy observation of $v = u + v_\sigma$ (where v_σ is a realization of ξ_σ) with the elements $w_1, \ldots, w_n$ obtained via

$$w_i = K_i K^{-1} v = K_i f,$$

that is, the integration

$$w_i = K_i f. \tag{4.5.16}$$

Figure 4.10 illustrates a three-mode $m = 3$ noisy signal, the correct determination of $n = m = 3$, and the recovery of its modes. Figure 4.10.1 displays the total observed signal $v = u + v_\sigma$, and the three modes v_1, v_2, and v_3 constituting $u = v_1 + v_2 + v_3$ are displayed in Fig. 4.10.5, 6, and 7, along with their recoveries w_1, w_2, and w_3.[2] Figure 4.10.8 also shows approximations of the instantaneous frequencies obtained as

$$w_{i,E}(t) := \text{argmax}_{\omega:(t, \omega) \in A(i)} E(t, \omega). \tag{4.5.17}$$

---

[2]The recoveries w_i in Fig. 4.10.5, 6, and 7 are indicated in red and the modes v_i of the signal are in blue. When the recovery is accurate, the red recovery blocks the blue and appears red.

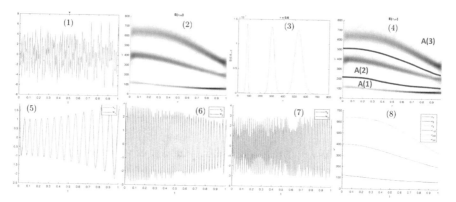

Figure 4.10: (1) The signal $v = u + v_\sigma$ where $u = v_1 + v_2 + v_3$, $v_\sigma \sim \mathcal{N}(0, \sigma^2 \delta(t-s))$ and $\sigma = 0.01$, (2) $(\tau, \omega) \to E(\tau, \omega)$ defined by (4.5.13) (one can identify three stripes), (3) $\omega \to E(0.6, \omega)$, (4) partitioning $[0, 1] \times [\omega_{\min}, \omega_{\max}] = \cup_{i=1}^3 A(i)$ of the time frequency domain into three disjoint subsets identified from E, (5) v_1 and its approximation w_1, (6) v_2 and its approximation w_2, (7) v_3 and its approximation w_3, and (8) ω_1, ω_2, and ω_3 and their approximations $\omega_{1,E}, \omega_{2,E}$, and $\omega_{3,E}$.

4.6 Convergence of the Numerical Methods

This section, which can be skipped on the first reading, provides a rough overview of how the empirical approach described in Remark 4.5.1 generates convergence results.

The Reproducing Kernel Hilbert Space and the Square Root Operator   To keep this discussion simple, we assume that the reproducing kernel K is continuous, and its corresponding integral operator $\bar{K}$ is injective (the more general case is handled by quotienting with respect to its nullspace). Then, the RKHS H_K can be described as the image $\bar{K}^{\frac{1}{2}}(L^2[0, 1]) \subset L^2[0, 1]$ of the unique positive symmetric square root of $\bar{K}$, and the map $\bar{K}^{\frac{1}{2}} : L^2[0, 1] \to H_K$ is an isometric isomorphism, see e.g., [62, Thm. 17.12].

The Gaussian Process and Least-Squares Collocation   By the zero–one law of Lukić and Beder [70, Thm. 7.2], the Gaussian stochastic process with covariance K has its sample paths in H_K with probability 1. Consequently, the Gaussian stochastic process with covariance K will have some approximation error when the observation v is not an element of H_K. This is the classical situation justifying the employment of Tikhonov regularization, motivating our introduction of the additive white noise component to the stochastic model. However, before we discuss Tikhonov regularization, let us begin with the case when v is an element of H_K. Then, according to Engl, Hanke, and Neubauer's

[31, Ex. 3.25] analysis of the least-squares collocation method in [31, Ex. 3.25] applied to solving the operator equation $\bar{K}^{\frac{1}{2}} f = v$, where $\bar{K}^{\frac{1}{2}}$ is considered as $\bar{K}^{\frac{1}{2}} : L^2[0,1] \to H_K$, application of the *dual least-squares* method of regularization, described in Engl, Hanke, and Neubauer [31, Ch. 3.3], reveals that our collocation discretization produces the least-squares collocation approximation f_m of the solution f of $\bar{K}^{\frac{1}{2}} f = v$, i.e., the minimal norm solution f_m of $Q_m \bar{K}^{\frac{1}{2}} f_m = Q_m v$, where $Q_m : H_K \to H_K$ denotes the H_K-orthogonal projection onto the span $\mathcal{Y}_m$ of the representers $\Phi_{x_j} \in H_K$ of the point evaluations at the collocation points x_j (i.e., we have $\langle w, \Phi_{x_j} \rangle_{H_K} = w(x_j)$, $w \in H_K$, $j = 1, \ldots, m$).

Tikhonov Regularization   Engl et al. [31, Thm. 3.24] assert that the resulting solution f_m satisfies $f_m = P_m f$, where $P_m : L^2[0,1] \to L^2[0,1]$ is the orthogonal projection onto $\bar{K}^{\frac{1}{2},*} \mathcal{Y}_m$. Quantitative analysis of the convergence of f_m to f is then a function of the strong convergence of P_m to the identity operator and can be assessed in terms of the expressivity of the set of representers Φ_{x_j}. For v not an element of H_K, Tikhonov regularization is applied together with least-squares collocation as in [31, Ch. 5.2].

Chapter 5

Additional Programming Modules and Squeezing

In this chapter, we continue the development of KMDNets by introducing additional programming modules and design them to develop flexible versions of synchrosqueezing, due to Daubechies et al. [21, 22]. The KMDNets described in Chap. 4 not only introduce hierarchical structures to implement nonlinear estimations using linear techniques, but can also be thought of as a sparsification technique whose goal is to reduce the computational complexity of solving the corresponding GPR problem, much like the sparse methods have been invented for GPR discussed in Sect. 2.2. The primary difference is that, whereas those methods generally use a set of inducing points determining a low-rank approximation and then choose the location of those points to optimize its approximation, here we utilize the landscape of the energy function $E : \mathcal{I} \to \mathbb{R}_+$, defined in (4.1.1) and analyzed in Proposition 4.1.1, interpreted as *alignment energies* near (4.1.4). In this section, this analog of sparse methods will be further developed for the KMDNets using the energy alignment landscape to further develop programming modules that improve the efficacy and accuracy of the reconstruction. For another application of the alignment energies in model construction, see Hamzi and Owhadi [43, Sec. 3.3.2] where it is used to estimate the optimal time lag of a ARMA-like time series model.

In the approach described in Sect. 4.4, $\mathcal{I}^{(k)}$ was partitioned into subsets $(j^{(k)})_{j \in \mathcal{I}^{(k+1)}}$ and the $Q_i^{(k)}$ were integrated (that is, summed over or average-pooled) using (4.3.1) and (4.3.15) in Line 4 of Algorithm 1, over each subset to obtain the $Q_j^{(k+1)}$. This partitioning approach can naturally be generalized to a *domain decomposition* approach by letting the subsets be non-disjoint and

H. Owhadi et al., *Kernel Mode Decomposition and the Programming of Kernels*, Surveys and Tutorials in the Applied Mathematical Sciences 8, https://doi.org/10.1007/978-3-030-82171-5_5

such that, for some k, $\cup_{j\in\mathcal{I}^{(k+1)}} j^{(k)}$ forms a strict subset[1] of $\mathcal{I}^{(k)}$ (i.e., some $i \in \mathcal{I}^{(k)}$ may not have descendants). We will now generalize the relation $\rightsquigarrow$ so as to (1) not satisfy Condition 4.3.2, that is, it does not define a map (a label i may have multiple descendants) (2) be non-directed, that is, not satisfy Definition 4.3.1 (some $j \in \mathcal{I}^{(k+1)}$ may have descendants in $\mathcal{I}^{(k)}$) and (3) enable loops.

With this generalization, the proposed framework is closer (in spirit) to an object-oriented programming language than to a meta-algorithm. This is consistent with what Yann LeCun in his recent lecture at the SIAM Conference on Mathematics of Data Science (MDS20) [64] has stated; paraphrasing him: "The types of architectures people use nowadays are not just chains of alternating linear and pointwise nonlinearities, they are more like programs now." We will therefore describe it as such via the introduction of additional elementary programming modules and illustrate the proposed language by programming increasingly efficient networks for mode decomposition.

5.1 Elementary Programming Modules

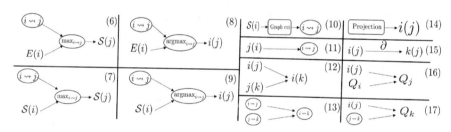

Figure 5.1: Elementary programming modules.

We will now introduce new elementary programming modules in addition to the five illustrated in Fig. 4.4 and discussed in Sect. 4.2. These new modules are illustrated in Fig. 5.1, beginning with module (6). Here they will be discussed abstractly but forward reference to specific examples. The first module (module (6)) of Fig. 5.1 replaces the average-pooling operation to define the energy E by a max-pool operation. More precisely module (6) combines a relation $i \rightsquigarrow j$ with an energy E to produce a *max-pool energy* via

$$\mathcal{S}(j) = \max_{i \rightsquigarrow j} E(i), \qquad (5.1.1)$$

where $i \rightsquigarrow j$ here is over i from the previous level to that of j. In what follows we will adhere to this semantic convention. As shown in module (7), this

---

[1]Although the results of Theorem 4.3.3 do not hold true under this general domain decomposition, those of Theorem 3.1.3 remain true between levels k and q (in particular, at each level k the $v_i^{(k)}$ are optimal recovered modes given the $Q_i^{(k)}$ and the observation v).

combination can also be performed starting with a *max-pool energy*, i.e., module (7) combines a relation $i \rightsquigarrow j$ with a max-pool energy $\mathcal{S}$ at one level to produce a *max-pool energy* at the next level via

$$\mathcal{S}(j) = \max_{i \rightsquigarrow j} \mathcal{S}(i) . \qquad (5.1.2)$$

Maximizers can naturally be derived from this max-pooling operation and modules (8) and (9) define $i(j)$ as the maximizer (or the set of maximizers if non-unique) of the energy or the max-pool energy. More precisely module (8) combines a relation $i \rightsquigarrow j$ with an energy function $E(i)$ to produce

$$i(j) = \mathrm{argmax}_{i \rightsquigarrow j}\, E(i) , \qquad (5.1.3)$$

and module (9)[2] combines a relation $i \rightsquigarrow j$ with a max-pool energy function $\mathcal{S}(i)$ to produce

$$i(j) = \mathrm{argmax}_{i \rightsquigarrow j}\, \mathcal{S}(i) . \qquad (5.1.4)$$

5.2   Programming the Network

Programming of the network is achieved by assembling the modules of Figs. 4.4 and 5.1 in a manner that (1) v is one of the inputs of the network and (if the network is used for mode decomposition/pattern recognition) (2) the modes v_m are one of the outputs of the network. As with any interpretable programming language avoiding inefficient coding and bugs remains important. We will use this language to program KMDNets.

---

[2]The description of the remaining modules (10)–(17), which can be skipped on first reading, is as follows. Similarly to module (3) of Fig. 4.4, module (10) of Fig. 5.1 combines the max-pool energy $\mathcal{S}$ with a graph operation to produce the ancestor-descendant relation $i \rightsquigarrow j$. We will show that module (10) leads to a more robust domain decomposition than module (3) due to its insensitivity to domain discretization. Module (11) uses the functional dependence $j(i)$ to define the relation $i \rightsquigarrow j$. Module (12) expresses the transitivity of function dependence, i.e., it combines $j(i)$ and $k(j)$ to produce $k(i)$. Similarly, module (13) expresses the transitivity of the relation $\rightsquigarrow$, i.e., $i \rightsquigarrow j$ and $j \rightsquigarrow k$ can be combined to produce $i \rightsquigarrow k$. Module (14) (analogously to module (4)) uses an injection step to define a functional dependence $i(j)$ (e.g., for the time–frequency application in Fig. 5.5, if $\mathcal{J}$ is the set of (τ, ω') and $\mathcal{I}$ is that of (τ, ω) the injection $\iota : \mathcal{I} \cap \mathcal{J} \to \mathcal{I}$ defines a functional dependence $i(j)$). Module (15) uses a functional dependence $i(j)$ to produce another functional dependence $k(j)$ (e.g., for the time–frequency–phase application in Figs. 5.7 and 5.8, we can define the functional dependence $(\tau, \omega')(\tau, \omega)$ from the functional dependence $(\tau, \omega, \theta)(\tau, \omega)$ via $\omega'(\tau, \omega) = \partial_\tau \theta(\tau, \omega)$). Module (16) utilizes the functional dependence $i(j)$ to produce a pullback covariance operator $Q_j := Q_{i(j)}$ ($:= \sum_{i \in i(j)} Q_i$ if $i(j)$ is a set-valued rather than a single-valued mapping). Module (17) combines a functional dependence $i(j)$ with a relation $j \rightsquigarrow k$ to produce a covariance operator Q_k (e.g., for the time–frequency–phase application of Figs. 5.7 and 5.8, for $i = (\tau, \omega, \theta) \in \mathcal{I}^{(1)}$ and $j = (\tau, k) \in \mathcal{I}^{(4)}$ where the index k is the mode index, the functional dependence $i(j)$ defines through (5.5.1) estimated phases $\theta_{k,e}(\cdot)$ which can then be substituted for $\theta(\cdot)$ in the kernel $K(s, t) = e^{-|s-t|^2/\gamma^2}\left(\cos(\theta(s))\cos(\theta(t)) + \sin(\theta(s))\sin(\theta(t))\right)$, producing for each mode index k a kernel with corresponding operator Q_k).

5.3   Alignments Calculated in L^2

We will now demonstrate both theoretically and experimentally that the alignment energy calculations may be simplified without loss in accuracy by computing them in L^2. Indeed, the calculation of the energies for our prototypical application was done with respect to the inner product defined by the inverse of the operator associated with K defined in (4.5.6), i.e., the energy of the mode (τ, ω, θ) was defined as $E(\tau, \omega, \theta) = v^T K^{-1} K_{\tau, \omega, \theta} K^{-1} v$ with $K_{\tau, \omega, \theta}$ defined in (4.5.3). The computational complexity of the method can be accelerated by (1) using the L^2 inner product instead of the one defined by K^{-1} (i.e., defining the energy of the mode (τ, ω, θ) by $E_2(\tau, \omega, \theta) = v^T K_{\tau, \omega, \theta} v$ (2) localizing this calculation in a time-window centered around τ and of width proportional to $1/\omega$.

Our experiments show that this simplification lowers the computational complexity of the proposed approach without impacting its accuracy. Three points justify this observation: (1) Replacing E by E_2 is equivalent to calculating mean-squared alignments with respect to the L^2-scalar product instead of the one induced by the inverse of the operator defined by K (2) In the limit where $\sigma \to \infty$ we have $E \approx \sigma^{-4} E_2$, therefore E and E_2 are proportional to each other in the high noise regime (3) If $\omega_{\min} = 0$ and $\omega_{\max} = \infty$, then K_u defined by (4.5.4) is the identity operator on L^2. We will now rigorously show that point (3) holds true when we extend the τ domain from $[0, 1]$ to $\mathbb{R}$ and when the base waveform is trigonometric, and then show in Sect. 9.1 that this result holds true independently of the base waveform being used.

Let us recall the Schwartz class of test functions

$$\mathcal{S} := \{f \in C^\infty(\mathbb{R}) : \sup_{x \in \mathbb{R}} |x^{m_1} D^{m_2} f(x)| < \infty, m_1, m_2 \in \mathbf{N}\}$$

and the confluent hypergeometric function $_1F_1$, defined by

$$_1F_1(\alpha, \gamma; z) = 1 + \frac{\alpha}{\gamma} \frac{z}{1!} + \frac{\alpha(\alpha + 1)}{\gamma(\gamma + 1)} \frac{z^2}{2!} + \frac{\alpha(\alpha + 1)(\alpha + 2)}{\gamma(\gamma + 1)(\gamma + 2)} \frac{z^3}{3!} + \cdots,$$

see, e.g., see Gradshteyn and Ryzhik [41, Sec. 9.21].

Theorem 5.3.1. *Consider extending the definition* (4.5.4) *of the kernel* K_u *so that the range of* ω *is extended from* $[\omega_{\min}, \omega_{\max}]$ *to* $\mathbb{R}_+$ *and that of* τ *is extended from* $[0, 1]$ *to* $\mathbb{R}$*, so that*

$$K_\beta(s, t) = \int_{-\pi}^{\pi} \int_{\mathbb{R}_+} \int_{\mathbb{R}} K_{\tau, \omega, \theta}(s, t) d\tau \, d\omega \, d\theta, \quad s, t \in \mathbb{R},$$

where, as before,

$$K_{\tau, \omega, \theta}(s, t) := \chi_{\tau, \omega, \theta}(s) \chi_{\tau, \omega, \theta}(t), \quad s, t \in \mathbb{R},$$

but where we have introduced a perturbation parameter $0 \leqslant \beta \leqslant 1$ *defining the Gabor wavelets*

$$\chi_{\tau, \omega, \theta}(t) := \left(\frac{2}{\alpha^2 \pi^3}\right)^{\frac{1}{4}} \omega^{\frac{1 - \beta}{2}} \cos\big(\omega(t - \tau) + \theta\big) e^{-\frac{\omega^2 (t - \tau)^2}{\alpha^2}}, \qquad t \in \mathbb{R}, \quad (5.3.1)$$

defining the elementary kernels. Defining the scaling constant

$$H(\beta) := 2^{\beta-1}\sqrt{\pi}(\sqrt{2}\alpha)^{1-\beta}\Gamma(\tfrac{\beta}{2}))e^{-\frac{\alpha^2}{2}}{}_1F_1\left(\frac{\beta}{2},\frac{1}{2};\frac{\alpha^2}{2}\right),$$

let $\mathcal{K}_\beta$ denote the integral operator

$$(\mathcal{K}_\beta f)(s) := \frac{1}{H(\beta)}\int_{\mathbb{R}} K_\beta(s,t)f(t)dt$$

associated with the kernel K_β scaled by $H(\beta)$. Then we have the semigroup property

$$\mathcal{K}_{\beta_1}\mathcal{K}_{\beta_2}f = \mathcal{K}_{\beta_1+\beta_2}f, \quad f \in \mathcal{S}, \quad \beta_1,\beta_2 > 0, \beta_1 + \beta_2 < 1,$$

and

$$\lim_{\beta\to 0}(\mathcal{K}_\beta f)(x) = f(x), \quad x \in \mathbb{R}, \quad f \in \mathcal{S},$$

where the limit is taken from above.

5.4 Squeezing

We will now present an interpretation and a variant (illustrated in Fig. 5.3) of the synchrosqueezing transform due Daubechies et al. [21, 22] (see Sect. 2.4 for a description), in the setting of KMDNets, and thereby initiate its GP regression version. We will demonstrate that this version generalizes to the case where the basic waveform is non-periodic and/or unknown. We use the setting and notations of Sect. 4.5.

Let f be the solution of $Kf = v$ (4.5.12) and let

$$E(\tau,\omega,\theta) := \int_0^1\int_0^1 f(s)K_{\tau,\omega,\theta}(s,t)f(t)\,ds\,dt \qquad (5.4.1)$$

be the energy of the mode indexed by (τ,ω,θ). For $(\tau,\omega) \in [0,1]\times[\omega_{\min},\omega_{\max}]$, write

$$\theta_e(\tau,\omega) := \operatorname{argmax}_{\theta\in(-\pi,\pi]} E(\tau,\omega,\theta). \qquad (5.4.2)$$

Since the definitions (4.5.1) of $\chi_{\tau,\omega,\theta}$ and (4.5.9) of $\chi_{\tau,\omega,c}$ and $\chi_{\tau,\omega,s}$, together with the cosine summation formula, imply that

$$\chi_{\tau,\omega,\theta}(t) = \frac{1}{\sqrt{\pi}}(\chi_{\tau,\omega,c}(t)\cos(\theta) - \chi_{\tau,\omega,s}(t)\sin(\theta)), \quad t \in \mathbb{R},$$

it follows that, if we define

$$W_c(\tau,\omega) := \int_0^1 \chi_{\tau,\omega,c}(t)f(t)\,dt$$

$$W_s(\tau,\omega) := \int_0^1 \chi_{\tau,\omega,s}(t)f(t)\,dt, \qquad (5.4.3)$$

we obtain

$$\int_0^1 \chi_{\tau,\omega,\theta}(t)f(t)\,dt = \frac{1}{\sqrt{\pi}}\left(\cos(\theta)W_c(\tau,\omega) - \sin(\theta)W_s(\tau,\omega)\right). \qquad (5.4.4)$$

Consequently, we deduce from (5.4.1) and (4.5.3) that

$$E(\tau,\omega,\theta) = \frac{1}{\pi}\left(\cos(\theta)W_c(\tau,\omega) - \sin(\theta)W_s(\tau,\omega)\right)^2. \qquad (5.4.5)$$

It follows, when either $W_c(\tau,\omega) \neq 0$ or $W_s(\tau,\omega) \neq 0$, that

$$\theta_e(\tau,\omega) = \text{phase}\left(W_c(\tau,\omega) - iW_s(\tau,\omega)\right), \qquad (5.4.6)$$

where, for a complex number z,

$$\text{phase}(z) := \theta \in (-\pi,\pi] : z = re^{i\theta},\ r > 0. \qquad (5.4.7)$$

Moreover, it follows from (4.5.8), (4.5.13), and (5.4.1) that

$$E(\tau,\omega) = \int_{-\pi}^{\pi} E(\tau,\omega,\theta)d\theta,$$

so that it follows from (5.4.5) that

$$E(\tau,\omega) = W_c^2(\tau,\omega) + W_s^2(\tau,\omega). \qquad (5.4.8)$$

Now consider the mode decomposition problem with observation $v = \sum v_i$ under the assumption that the phases vary much faster than the amplitudes. It follows that for the determination of frequencies (not the determination of the phases) we can, without loss of generality, assume each mode is of the form

$$v_i(t) = a_i(t)\cos(\theta_i(t)), \qquad (5.4.9)$$

where a_i is slowly varying compared to θ_i. We will use the symbol $\approx$ to denote an informal approximation analysis. Theorem 5.3.1 asserts that K is approximately a multiple of the identity operator, so we conclude that the solution f to $Kf = v$ in (4.5.12) is $f \approx cv$ for some constant c. Because we will be performing a phase calculation the constant c is irrelevant and so can be set to 1, that is we have $f \approx v$ and therefore we can write (5.4.3) as

$$W_c(\tau,\omega) \approx \int_0^1 \chi_{\tau,\omega,c}(t)v(t)\,dt$$

$$W_s(\tau,\omega) \approx \int_0^1 \chi_{\tau,\omega,s}(t)v(t)\,dt. \qquad (5.4.10)$$

For fixed τ, for t near τ,

$$v_i(t) \approx a_i(\tau)\cos((t-\tau)\dot{\theta}_i(\tau) + \theta_i(\tau)) \qquad (5.4.11)$$

so that, since the frequencies $\dot{\theta}_i$ are relatively large and well separated, it follows from the nullification effect of integrating cosines of high frequencies, that for $\omega \approx \dot{\theta}_i(\tau)$, (5.4.10) holds true with v_i instead of v in the right-hand side. Because the amplitudes of v_i in (5.4.9) are slowly varying compared to their frequencies, it again follows from the nullification effect of integrating cosines of high frequencies, the approximation formula (5.4.11), the representation (4.5.9) of $\chi_{\tau,\omega,c}$ and $\chi_{\tau,\omega,s}$ and the sine and cosine summation formulas, that

$$W_c(\tau,\omega) \approx a_i(\tau)\cos(\theta_i(\tau))\int_0^1 \chi_{\tau,\omega,c}(t)\cos((t-\tau)\omega)\,dt$$

$$W_s(\tau,\omega) \approx -a_i(\tau)\sin(\theta_i(\tau))\int_0^1 \chi_{\tau,\omega,s}(t)\sin((t-\tau)\omega)\,dt\,.$$

Since the representation (4.5.9) of $\chi_{\tau,\omega,c}$ and $\chi_{\tau,\omega,s}$, and the sine and cosine summation formulas, also imply that $\int_0^1 \chi_{\tau,\omega,c}(t)\cos((t-\tau)\omega)\,dt \approx \int_0^1 \chi_{\tau,\omega,s}(t)\sin((t-\tau)\omega)\,dt > 0$, it follows that

$$W_c(\tau,\omega) - iW_s(\tau,\omega) \approx a_i(\tau)e^{i\theta_i(\tau)}\int_0^1 \chi_{\tau,\omega,c}(t)\cos((t-\tau)\omega)\,dt\,,$$

so that $\theta_e(\tau,\omega)$, defined in (5.4.6), is an approximation of $\theta_i(\tau)$, and

$$\omega_e(\tau,\omega) = \frac{\partial\theta_e}{\partial\tau}(\tau,\omega) \tag{5.4.12}$$

is an approximation of the instantaneous frequency $\dot{\theta}_i(\tau)$.

Remark 5.4.1. *In the discrete case, on a set $\{\tau_k\}$ of points, we proceed differently than in (5.4.12). Ignoring for the moment the requirement (5.4.7) that the phase $\theta_e(\tau,\omega)$ defined in (5.4.6) lies in $(-\pi, \pi]$, an accurate finite difference approximation $\omega_e(\tau_k,\omega)$ to the frequency is determined by*

$$\theta_e(\tau_k,\omega) + \omega_e(\tau_k,\omega)(\tau_{k+1}-\tau_k) = \theta_e(\tau_{k+1},\omega).$$

To incorporating the requirement, it is natural to instead define $\omega_e(\tau_k,\omega)$ as solving

$$e^{i\,\omega_e(\tau_k,\omega)(\tau_{k+1}-\tau_k)}e^{i\theta_e(\tau_k,\omega)} = e^{i\theta_e(\tau_{k+1},\omega)},$$

which using (5.4.6) becomes

$$e^{i\,\omega_e(\tau_k,\omega)(\tau_{k+1}-\tau_k)}e^{i\,\text{phase}(W_c(\tau_k,\omega)-iW_s(\tau_k,\omega))} = e^{i\,\text{phase}(W_c(\tau_{k+1},\omega)-iW_s(\tau_{k+1},\omega))},$$

and has the solution

$$\omega_e(\tau_k,\omega) = \frac{1}{\tau_{k+1}-\tau_k}\,\text{atan2}\left(\frac{W_c(\tau_{k+1},\omega)W_s(\tau_k,\omega) - W_s(\tau_{k+1},\omega)W_c(\tau_k,\omega)}{W_c(\tau_{k+1},\omega)W_c(\tau_k,\omega) + W_s(\tau_{k+1},\omega)W_s(\tau_k,\omega)}\right),$$
$$\tag{5.4.13}$$

where atan2 is Fortran's four-quadrant inverse tangent.

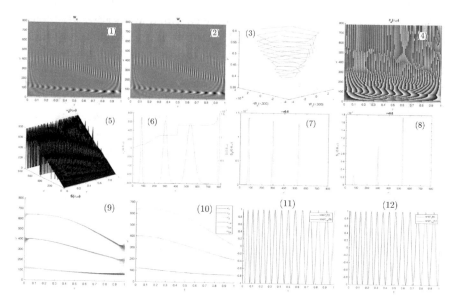

Figure 5.2: (1) $W_c(\tau, \omega)$, (2) $W_s(\tau, \omega)$, (3) $\tau \to (W_c(\tau, 300), W_s(\tau, 300), \tau)$, (4) $(\tau, \omega) \to \theta_e(\tau, \omega)$, (5) $(\tau, \omega) \to \omega_e(\tau, \omega)$, (6) $\omega \to \omega_e(0.6, \omega)$ and $\omega \to E(0.6, \omega)$, (7) $\omega \to \mathcal{S}(0.6, \omega)$, (8) $\omega \to \mathcal{S}_E(0.6, \omega)$, (9) $(\tau, \omega) \to \mathcal{S}(\tau, \omega)$, (10) $t \to \omega_i(t)$ and $t \to \omega_{i,e}(t)$ for $i \in \{1, 2, 3\}$, (11) $t \to \cos(\theta_1(t))$ and $t \to \cos(\theta_{1,e}(t))$, (12) $t \to \sin(\theta_1(t))$ and $t \to \sin(\theta_{1,e}(t))$.

In preparation for illustrating the application of the programming of KMD-Nets, as a synchrosqueezing algorithm, to the decomposition problem when v and its modes are as in Fig. 4.10, Fig. 5.2 illustrates the basic quantities we have just been developing. In particular,

- The functions W_c and W_s are shown in Fig. 5.2.1, 2.

- The function $\tau \to (W_c(\tau, 300), -W_s(\tau, 300))$ is shown in Fig. 5.2.3 with τ the vertical axis. The functions $\theta_e(\tau, 300)$, $E(\tau, 300)$, and $\omega_e(\tau, 300)$ are the phase, square modulus, and angular velocity of this function.

- The functions $(\tau, \omega) \to \theta_e(\tau, \omega)$, $\tau \to \theta_e(\tau, \omega_{i,E}(\tau))$ (with $\omega_{i,E}$ defined in (4.5.17)) and $t \to \theta_i(t)$ are shown in Fig. 5.2.4, 11, and 12. Observe that $\tau \to \theta_e(\tau, \omega_{i,E}(\tau))$ is an approximation of $\tau \to \theta_i(\tau)$.

- The functions $(\tau, \omega) \to \omega_e(\tau, \omega)$, $\omega \to \omega_e(0.6, \omega)$ and $\tau \to \omega_e(\tau, \omega_{i,E}(\tau))$ are shown in Fi. 5.2.5, 6, and 10. Observe that $\tau \to \omega_e(\tau, \omega_{i,E}(\tau))$ is an approximation of the instantaneous frequency $\tau \to \omega_i(\tau) = \dot\theta_i(\tau)$ of the mode v_i.

To describe the remaining components of Fig. 5.2 and simultaneously complete the application of the programming of KMDNets as a synchrosqueezing algorithm and introduce a *max-pool* version of synchrosqueezing, we now introduce the synchrosqueezed energy $\mathcal{S}_E(\tau, \omega)$ and the max-pool energy $\mathcal{S}(\tau, \omega)$: Motivated by the synchrosqueezed transform introduced in Daubechies et al. [21], the synchrosqueezed energy $\mathcal{S}_E(\tau, \omega)$ is obtained by transporting the energy $E(\tau, \omega)$ via the map $(\tau, \omega) \to (\tau, \omega_e(\tau, \omega))$ (as discussed in Sect. 2.4, especially near (2.4.3)) and therefore satisfies

$$\int_{\omega_{\min}}^{\omega_{\max}} \varphi(\omega) \mathcal{S}_E(\tau, \omega)\, d\omega = \int_{\omega_{\min}}^{\omega_{\max}} \varphi(\omega_e(\tau, \omega')) E(\tau, \omega')\, d\omega'$$

for all regular test function φ, i.e.,

$$\mathcal{S}_E(\tau, \omega) = \lim_{\delta \to 0} \frac{1}{\delta} \int_{\omega' : \omega \leqslant \omega_e(\tau, \omega') \leqslant \omega + \delta} E(\tau, \omega')\, d\omega', \qquad (5.4.14)$$

where we numerically approximate (5.4.14) by taking δ small.

Returning to the application, the transport of the energy $E(\tau, \omega)$ via the map $(\tau, \omega) \to (\tau, \omega_e(\tau, \omega))$ is illustrated for $\tau = 0.6$ by comparing the plots of the functions $\omega \to \omega_e(0.6, \omega)$ and $\omega \to E(0.6, \omega)$ in Fig. 5.2.6 with the function $\omega \to \mathcal{S}_E(0.6, \omega)$ shown in Fig. 5.2.8. As in [21], the value of $\mathcal{S}_E(\tau, \omega)$ (and thereby the height of the peaks in Fig. 5.2.8) depends on the discretization and the measure $d\omega'$ used in the integration (5.4.14). For example, using a logarithmic discretization or replacing the Lebesgue measure $d\omega'$ by $\omega' d\omega'$ in (5.4.14) will impact the height of those peaks. To avoid this dependence on the choice of measure, we define the max-pool energy

$$\mathcal{S}(\tau, \omega) = \max_{\omega' : \omega_e(\tau, \omega') = \omega} E(\tau, \omega'), \qquad (5.4.15)$$

illustrated in Fig. 5.2.9. Comparing Fig. 5.2.6, 7, and 8, observe that, although both synchrosqueezing and max-pooling decrease the width of the peaks of the energy plot $\omega \to E(0.6, \omega)$, only max-squeezing preserves their heights (as noted in [21, Sec. 2] a discretization dependent weighting of $d\omega'$ would have to be introduced to avoid this dependence).

Figure 5.3 provides an interpretation of the synchrosqueezed and max-pool energies $\mathcal{S}_E(\tau, \omega)$ and $\mathcal{S}(\tau, \omega)$ in the setting of KMDNet programming, where we

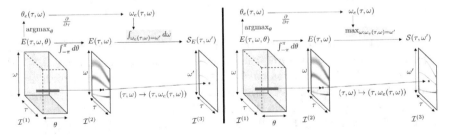

Figure 5.3: Synchrosqueezed (left) and max-pool (right) energies.

note that the left (synchrosqueezed) and right (max-pool) sub-figures are identical except for the highlighted portions near their top center. In that interpretation $\mathcal{I}^{(1)}$ and $\mathcal{I}^{(2)}$ are, as in Sect. 4.5 and modulo the noise mode σ, respectively, the set of time–frequency–phase labels $(\tau, \omega, \theta) \in [0, 1] \times [\omega_{\min}, \omega_{\max}] \times (-\pi, \pi]$ and the set of time–frequency labels $(\tau, \omega) \in [0, 1] \times [\omega_{\min}, \omega_{\max}]$. Modulo the noise label σ, $\mathcal{I}^{(3)}$ is the range of $(\tau, \omega) \to (\tau, \omega_e(\tau, \omega))$ and the ancestors of $(\tau, \omega') \in \mathcal{I}^{(3)}$ are the (τ, ω) such that $\omega' = \omega_e(\tau, \omega)$. Then, in that interpretation, the synchrosqueezed energy is simply the level 3 energy $E^{(3)}$, whereas $\mathcal{S}(\tau, \omega)$ is the level 3 max-pool energy $\mathcal{S}^{(3)}$. Note that the proposed approach naturally generalizes to the case where the periodic waveform y is known and non-trigonometric by simply replacing the cosine function in (4.5.1) by y.

5.5   Crossing Instantaneous Frequencies

Let us now demonstrate the effectiveness of the max-pooling technique in its ability to perform mode recovery when the instantaneous frequencies of the modes cross. Consider the noisy signal v illustrated in Fig. 5.4.1. This signal

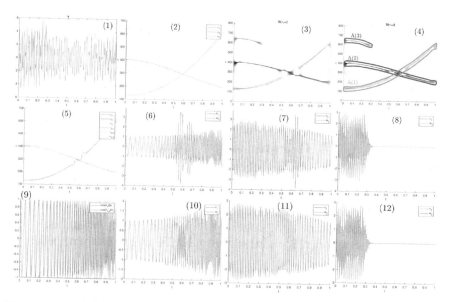

Figure 5.4: (1) The signal $v = v_1 + v_2 + v_3 + v_\sigma$ where $v_\sigma \sim \mathcal{N}(0, \sigma^2 \delta(s - t))$ and $\sigma = 0.01$, (2) instantaneous frequencies $t \to \omega_i(t)$ of the modes $i = 1, 2, 3$, (3) $(\tau, \omega) \to \mathcal{S}(\tau, \omega)$, (4) Sub-domains $A(1), A(2)$, and $A(3)$ of the time–frequency domain, (5) approximated instantaneous frequencies $t \to \omega_{i,e}(t)$ of the modes $i = 1, 2, 3$, (6, 7, 8) v_1, v_2, v_3 and their approximations w_1, w_2, w_3 obtained from the network shown in Fig. 5.5, (9) phase θ_1 and its approximation $\theta_{1,e}$, (10, 11, 12) v_1, v_2, v_3 and their approximations w_1, w_2, w_3 obtained from the network shown in Fig. 5.7.

is composed of 4 modes, $v = v_1 + v_2 + v_3 + v_\sigma$, where $v_\sigma \sim \mathcal{N}(0, \sigma^2 \delta(s-t))$ is a white noise realization with $\sigma = 0.01$. The modes v_1, v_2, v_3 are shown in Fig. 5.4.6, 7, and 8, and their instantaneous frequencies $\omega_1, \omega_2, \omega_3$ are shown in Fig. 5.4.2 (see Footnote 2). Note that ω_1 and ω_2 cross each other around $t \approx 0.6$ and v_3 vanishes around $t \approx 0.3$. We now program two KMDNets and describe their accuracy in recovering those modes.

The first network, illustrated in Figs. 5.5 and 5.6, recovers approximations to v_1, v_2, v_3 by identifying three subsets $A(1), A(2), A(3)$ of the time–frequency domain (τ, ω) and integrating the kernel $K_{\tau,\omega}$ (defined as in (4.5.8)) over those subsets (as in (4.5.15)). For this example, the subsets $A(1), A(2), A(3)$ are shown in Fig. 5.4.4 and identified as narrow sausages defined by the peaks of the max-pool energy $\mathcal{S}^{(3)}(\tau, \omega')$ (computed as in (5.4.15)) shown in 5.4.3. The corresponding approximations w_1, w_2, w_3 (obtained as in (4.5.16)) of the modes v_1, v_2, v_3 are shown in Fig. 5.4.6, 7, and 8. Note the increased approximation error around $t \approx 0.6$ corresponding to the crossing point between ω_1 and ω_2 and $A(1)$ and $A(2)$. The estimated instantaneous frequencies $\omega_{i,e}(\tau) = \omega_e\big(\tau, \text{argmax}_{\omega:(\tau,\omega)\in A(i)} \mathcal{S}^{(3)}(\tau, \omega)\big)$ illustrated in Fig. 5.4.5 also show an increased estimation error around that crossing point.

The second network, illustrated in Figs. 5.7 and 5.8, proposes a more robust approach based on the estimates $\theta_{i,e}$ of instantaneous phases θ_i obtained as

$$\theta_{i,e}(\tau) = \theta_e\big(\tau, \text{argmax}_{\omega:(\tau,\omega)\in A(i)} \mathcal{S}^{(3)}(\tau, \omega)\big), \qquad (5.5.1)$$

where the $A(i)$ are obtained as in the first network, illustrated in Fig. 5.5, and $\theta_e(\tau, \omega)$, used in the definition (5.5.1) of $\theta_{e,i}(\tau)$, is identified as in (5.4.6). To

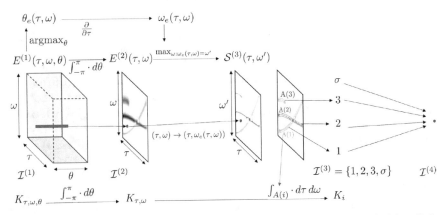

Figure 5.5: Recovery from domain decomposition. The left-hand side of the figure is that of the right-hand side (corresponding to max-pooling) of Fig. 5.3. The remaining part is obtained by identifying three subsets $A(1), A(2), A(3)$ of the time–frequency domain (τ, ω) and integrating the kernel $K_{\tau,\omega}$ (defined as in (4.5.8)) over those subsets (as in (4.5.15)).

recover the modes v_i, the proposed network proceeds as in Example 3.3.1 by introducing the kernels

$$K_i(s,t) = e^{-\frac{(t-s)^2}{\gamma^2}} \left(\cos(\theta_{i,e}(t)) \cos(\theta_{i,e}(s)) + \sin(\theta_{i,e}(t)) \sin(\theta_{i,e}(s)) \right), \quad (5.5.2)$$

with $\gamma = 0.2$. Defining K_σ as in (4.5.7), the approximations w_1, w_2, w_3 of the modes v_1, v_2, v_3, shown in Fig. 5.4.10, 11 and 12, are obtained as in (4.5.16) with f defined as the solution of $(K_1 + K_2 + K_3 + K_\sigma)f = v$. Note that the network illustrated in Fig. 5.7 can be interpreted as the concatenation of 2 networks. One aimed at estimating the instantaneous phases and the other aimed at recovering the modes based on those phases. This principle of network concatenation is evidently generic.

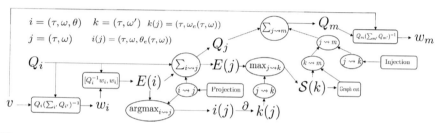

Figure 5.6: The KMDNet program corresponding to Fig. 5.5. Upper left provides the symbolic connections between the indices i, j, k and the time–frequency parameters along with the functional dependencies $i(j)$ and $k(j)$. Beginning with the input v in the lower left, the operators Q_i corresponding to the baby kernels $K_{\tau,\omega,\theta}$ are used to produce optimal recovery estimates w_i and the corresponding alignment energies $E(i)$. The projection function $j(i)$ taking (τ, ω, θ) to (τ, ω) is the relation $i \rightsquigarrow j$ which determines the integration operation $\int d\theta$ indicated as $\sum_{i \rightsquigarrow j}$ which then determines summed energies $E(j) := \sum_{i \rightsquigarrow j} E(i)$ and covariances $Q_j := \sum_{i \rightsquigarrow j} Q_i$. Moreover, the projection $i \rightsquigarrow j$ also determines a max operation $\arg\max_\theta$ which we denote by $\arg\max_{i \rightsquigarrow j}$ and the resulting function $\theta_e(\tau, \omega) := \arg\max_\theta E_{\tau,\omega,\theta}$, which determines the functional dependency $i(j) = (\tau, \omega, \theta_e(\tau, \omega))$. This function is then differentiated to obtain the functional relation $k(j) = (\tau, \omega_e(\tau, \omega))$ where $\omega_e(\tau, \omega) := \frac{\partial}{\partial \tau} \theta_e(\tau, \omega)$. This determines the relation $j \rightsquigarrow k$ which determines the maximization operation $\max_{j \rightsquigarrow k}$ that, when applied to the alignment energies $E(j)$, produces the max-pooled energies $\mathcal{S}(k)$. These energies are then used to determine a graph-cut establishing a relation $k \rightsquigarrow m$ where m is a mode index. Combining this relation with the injection $j \rightsquigarrow k$ determines the relation $j \rightsquigarrow m$ that then determines the summation $\sum_{j \rightsquigarrow m}$ over the preimages of the relation, thus determining operators Q_m indexed by the mode m by $Q_m := \sum_{j \rightsquigarrow m} Q_j$. Optimal recovery is then applied to obtain the estimates $w_m := Q_m (\sum_{m'} Q_{m'})^{-1}$.

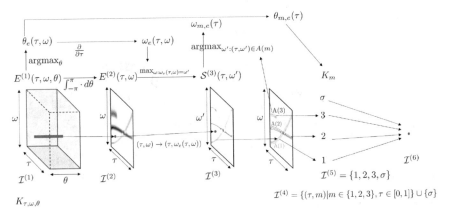

Figure 5.7: Recovery from instantaneous phases approximations. The left-hand side of the figure is that of the right-hand side (corresponding to max-pooling) of Fig. 5.3 and therefore also that of Fig. 5.5, and proceeding to the right as in Fig. 5.5, the three subsets $A(1), A(2), A(3)$ of the time–frequency domain (τ, ω) and integrating the kernel $K_{\tau,\omega}$ (defined as in (4.5.8)) over those subsets (as in (4.5.15)). However, to define the kernels K_m for the final optimal recovery, we define $\omega_{m,e}(\tau) := \arg\max_{\omega':(\tau,\omega')\in A(i)} S^3(\tau, \omega')$ to produce the θ function for each mode m through $\theta_{m,e}(\tau) = \theta_e(\tau, \omega_{m,e}(\tau))$. These functions are inserted into (5.5.2) to produce K_m and their associated operators Q_m which are then used in the finally recovery $w_m = Q_m(\sum_{m'} Q_{m'})^{-1}v$.

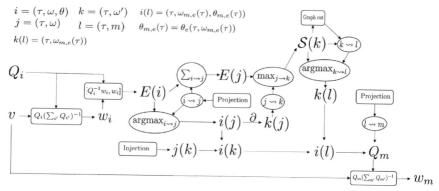

Figure 5.8: The KMDNet program corresponding to Fig. 5.7. Upper left provides the symbolic connections between the indices i, j, k, l and the time–frequency parameters along with the functional dependencies $i(l)$ and $k(l)$ and the definition of $\theta_{m,e}$. Beginning with the input v in the lower left, ignoring the bottom two rows for the moment, we begin very much as in Fig. 5.6 moving to the right until the determination of the energies $\mathcal{S}(k)$, the determination of a graph-cut and its resulting $k \rightsquigarrow l$, and the resulting $\arg\max$ relation $k(l) := \arg\max_{k \rightsquigarrow l} \mathcal{S}(k)$ which amounts to $k(l) = (\tau, \omega_{m,e}(\tau))$. Returning to the second row from the bottom, we compose the functional relations of the injection $j(k)$ and the $\arg\max$ function $i(j)$ determined by the relation $i \rightsquigarrow j$ and the energy $E(i)$, to obtain $i(k)$ and then compose this with the argmax function $k(l)$ to produce the functional dependence $i(l)$ defined by $i(l) = (\tau, \omega_{m,e}(\tau), \theta_{m,e}(\tau))$. Using the projection $l \rightsquigarrow m$, this determines the function $\theta_{m,e}(\cdot)$ corresponding to the mode label m. These functions are inserted into (5.5.2) to produce K_m and their associated operators Q_m which are then used in the finally recovery $w_m = Q_m(\sum_{m'} Q_{m'})^{-1} v$.

Chapter 6

Non-trigonometric Waveform and Iterated KMD

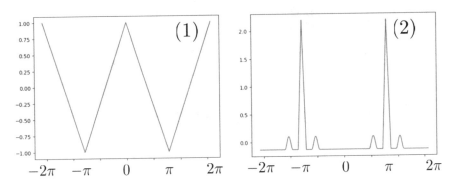

Figure 6.1: (1) Triangle base waveform, (2) EKG base waveform.

In this chapter we will consider the mode recovery Problem 1 generalized to the case where the base waveform of each mode is the same known, possibly non-trigonometric, square-integrable 2π-periodic function $t \to y(t)$, and develop an *iterated micro-local KMD* algorithm for its solution. The objective of this problem can be loosely expressed as solving the following generalization of Problem 1 towards the resolution of the more general Problem 2. We now switch the time domain from $[0, 1]$ to $[-1, 1]$.

Problem 4. *For $m \in \mathbb{N}^*$, let $a_1, \ldots, a_m$ be piecewise smooth functions on $[-1, 1]$, let $\theta_1, \ldots, \theta_m$ be strictly increasing functions on $[-1, 1]$, and let y be a square-integrable 2π-periodic function. Assume that m and a_i, θ_i are unknown,*

© The Author(s), under exclusive license to Springer Nature Switzerland AG 2021

H. Owhadi et al., *Kernel Mode Decomposition and the Programming of Kernels*, Surveys and Tutorials in the Applied Mathematical Sciences 8, https://doi.org/10.1007/978-3-030-82171-5_6

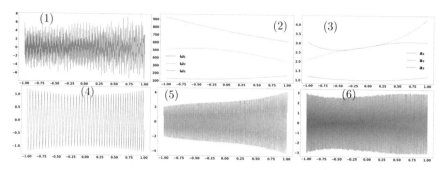

Figure 6.2:   Triangle base waveform: (1) Signal v, (2) Instantaneous frequencies $\omega_i := \dot{\theta}_i$, (3) Amplitudes a_i, (4, 5, 6) Modes v_1, v_2, v_3.

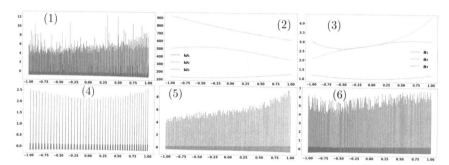

Figure 6.3: EKG base waveform: (1) Signal v, (2) Instantaneous frequencies $\omega_i := \dot{\theta}_i$, (3) Amplitudes a_i, (4, 5, 6) Modes v_1, v_2, v_3.

and the base waveform y is known. We further assume that $a_i(t) > \epsilon > 0$, and that $\dot{\theta}_i(t)/\dot{\theta}_j(t) \notin [1 - \epsilon, 1 + \epsilon]$ for all i, j, t. Given the observation $v(t) = \sum_{i=1}^m a_i(t)y(\theta_i(t))$ (for $t \in [-1, 1]$) recover the modes $v_i := a_i(t)y(\theta_i(t))$.

Example 6.0.1. *Figure 6.1 shows two full periods of two 2π-periodic base waveforms (triangle and EKG), which we will use in our numerical experiments/illustrations. The EKG (-like) waveform is $\big(y_{EKG}(t) - (2\pi)^{-1}$ $\int_0^{2\pi} y_{EKG}(s)\,ds\big)/\|y_{EKG}\|_{L^2([0,2\pi))}$ with $y_{EKG}(t)$ defined on $[0, 2\pi)$ as (1) $0.3 - |t - \pi|$ for $|t - \pi| < 0.3$ (2) $0.03\cos^2(\frac{\pi}{0.6}(t - \pi + 1))$ for $|t - \pi + 1| < 0.3$ (3) $0.03\cos^2(\frac{\pi}{0.6}(t - \pi - 1))$ for $|t - \pi - 1| < 0.3$ and (4) 0 otherwise.*

Our approach, summarized in Algorithm 2 and explained in the following sections, will be to (1) use the max-pool energy $\mathcal{S}$ (5.4.15) to obtain, using (5.5.1), an estimate of the phase $\theta_{\text{low}}(t)$ associated with the lowest instantaneous frequency $\omega_{\text{low}} = \dot{\theta}_{\text{low}}$ (as described in Sect. 6.2) (2) iterate a *micro-local*

KMD (presented in Sect. 6.1) of the signal v to obtain a highly accurate estimate of the phase/amplitude θ_i, a_i of their corresponding mode v_i (this iteration can achieve near machine-precision accuracies when the instantaneous frequencies are separated) (3) Peel off the mode v_i from v (4) iterate to obtain all the modes (5) perform a last micro-local KMD of the signal for higher accuracy. To illustrate this approach, in the next two sections, we will apply it to the signals v displayed in Figs. 6.2 and 6.3, where the modes of Fig. 6.2 are triangular, and those of Fig. 6.3 are EKG.

6.1 The Micro-Local KMD Module

We will now describe the micro-local KMD module, which will form the basis for the iterated micro-local KMD algorithm described in Sect. 6.3. It takes a time τ, an estimated phase function of i-th mode $\theta_{i,e}$, and a signal f, not necessarily equal to v. Suppose the i-th mode is of form $v_i(t) = a_i(t) y(\theta_i(t))$ and is indeed a mode within f. The module outputs, (1) an estimate $a(\tau, \theta_{i,e}, f)$ of the amplitude $a_i(\tau)$ of the mode v_i and (2) a correction $\delta\theta(\tau, \theta_{i,e}, f)$ determining an updated estimate $\theta_{i,e}(\tau) + \delta\theta(\tau, \theta_{i,e}, f)$ of the estimated mode phase function $\theta_{i,e}$. We assume that a_i is strictly positive, that is, $a_i(t) \geqslant a_0, t \in [-1, 1]$, for some $a_0 > 0$.

Indeed, given $\alpha > 0$, $\tau \in [-1, 1]$, differentiable strictly increasing functions θ_0 and θ_e on $[-1, 1]$, and $n \in \{0, \ldots, d\}$ (we set $d = 2$ in applications in this section), let $\chi_{n,c}^{\tau,\theta_e}$ and $\chi_{n,s}^{\tau,\theta_e}$ be the wavelets defined by

$$\chi_{n,c}^{\tau,\theta_e}(t) := \cos(\theta_e(t))(t - \tau)^n e^{-\left(\frac{\dot{\theta}_0(\tau)(t-\tau)}{\alpha}\right)^2}$$

$$\chi_{n,s}^{\tau,\theta_e}(t) := \sin(\theta_e(t))(t - \tau)^n e^{-\left(\frac{\dot{\theta}_0(\tau)(t-\tau)}{\alpha}\right)^2}, \tag{6.1.1}$$

and let ξ_{τ,θ_e} be the Gaussian process defined by

$$\xi_{\tau,\theta_e}(t) := \sum_{n=0}^{d} \left(X_{n,c} \chi_{n,c}^{\tau,\theta_e}(t) + X_{n,s} \chi_{n,s}^{\tau,\theta_e}(t) \right), \tag{6.1.2}$$

where $X_{n,c}, X_{n,s}$ are independent $\mathcal{N}(0, 1)$ random variables. The function θ_0 will be fixed throughout the iterations whereas the function θ_e will be updated. Let f_τ be the Gaussian windowed signal defined by

$$f_\tau(t) = e^{-\left(\frac{\dot{\theta}_0(\tau)(t-\tau)}{\alpha}\right)^2} f(t), \quad t \in [-1, 1], \tag{6.1.3}$$

and, for $(n, j) \in \{0, \ldots, d\} \times \{c, s\}$, let

$$Z_{n,j}(\tau, \theta_e, f) := \lim_{\sigma \downarrow 0} \mathbb{E}\left[X_{n,j} \big| \xi_{\tau,\theta_e} + \xi_\sigma = f_\tau \right], \tag{6.1.4}$$

where ξ_σ is white noise, independent of ξ_{τ,θ_e}, with variance σ^2. To compute $Z_{n,j}$, observe that since both ξ_{τ,θ_e} and ξ_σ are Gaussian fields, it follows from (3.3.1) that

$$\mathbb{E}\left[\xi_{\tau,\theta_e} \big| \xi_{\tau,\theta_e} + \xi_\sigma \right] = A_\sigma (\xi_{\tau,\theta_e} + \xi_\sigma)$$

for the linear mapping

$$A_\sigma = Q_{\tau,\theta_e}\left(Q_{\tau,\theta_e} + \sigma^2 I\right)^{-1},$$

where $Q_{\tau,\theta_e} : L^2 \to L^2$ is the covariance operator of the Gaussian field ξ_{τ,θ_e} and $\sigma^2 I$ is the covariance operator of ξ_σ. Using the characterization of the limit of Tikhonov regularization as the Moore-Penrose inverse, see, e.g., Barata and Hussein [7, Thm. 4.3], along with the orthogonal projections connected with the Moore-Penrose inverse, we conclude that $\lim_{\sigma\to 0} A_\sigma = P_{\chi^{\tau,\theta_e}}$, where $P_{\chi^{\tau,\theta_e}}$ is the L^2-orthogonal projection onto the span $\chi^{\tau,\theta_e} := \mathrm{span}\{\chi_{n,c}^{\tau,\theta_e}, \chi_{n,s}^{\tau,\theta_e}, n = 0,\ldots,d\}$, and therefore

$$\lim_{\sigma\to 0} \mathbb{E}\left[\xi_{\tau,\theta_e}\middle|\xi_{\tau,\theta_e} + \xi_\sigma\right] = P_{\chi^{\tau,\theta_e}}\left(\xi_{\tau,\theta_e} + \xi_\sigma\right). \qquad (6.1.5)$$

Since the definition (6.1.2) can be written $\xi_{\tau,\theta_e} = \sum_{n,j} X_{n,j} \chi_{n,j}^{\tau,\theta_e}$, summing (6.1.4) and using (6.1.5), we obtain

$$\sum_{n,j} Z_{n,j}(\tau,\theta_e,f)\chi_{n,j}^{\tau,\theta_e} = P_{\chi^{\tau,\theta_e}} f_\tau. \qquad (6.1.6)$$

Consider the vector function $Z(\tau,\theta_e,f) \in \mathbb{R}^{2d+2}$ with components $Z_{n,j}(\tau,\theta_e,f)$, the $2d+2$ dimensional Gaussian random vector X with components $X_{n,j}, (n,j) \in \{0,\ldots,d\} \times \{c,s\}$, and the $(2d+2) \times (2d+2)$ matrix A^{τ,θ_e} defined by

$$A^{\tau,\theta_e}_{(n,j),(n',j')} := \langle \chi_{n,j}^{\tau,\theta_e}, \chi_{n',j'}^{\tau,\theta_e}\rangle_{L^2[-1,1]}. \qquad (6.1.7)$$

Straightforward linear algebra along with (6.1.6) establishes that the vector $Z(\tau,\theta_e,f)$ can be computed as the solution of the linear system

$$A^{\tau,\theta_e} Z(\tau,\theta_e,f) = b^{\tau,\theta_e} f, \qquad (6.1.8)$$

where $b^{\tau,\theta_e}(f)$ is the $\mathbb{R}^{2d+2}$ vector with components $b_{n,j}^{\tau,\theta_e}(f) := \langle \chi_{n,j}^{\tau,\theta_e}, f_\tau\rangle_{L^2}$. See sub-figures (1) and (2) of both the top and bottom of Fig. 6.6 for illustrations of the windowed signal $f_\tau(t)$ and of its projection $\lim_{\sigma\downarrow 0} \mathbb{E}\left[\xi_{\tau,\theta_e}\middle|\xi_{\tau,\theta_e} + \xi_\sigma = f_\tau\right]$ in (6.1.5) corresponding to the signals f displayed in Figs. 6.2 and 6.3.

To apply these formulations to construct the module, suppose that f is a single mode

$$f(t) = a(t)\cos(\theta(t)),$$

so that

$$f_\tau(t) = e^{-\left(\frac{\dot\theta_0(\tau)(t-\tau)}{\alpha}\right)^2} a(t)\cos(\theta(t)), \qquad (6.1.9)$$

and consider the modified function

$$\bar{f}_\tau(t) = e^{-\left(\frac{\dot\theta_0(\tau)(t-\tau)}{\alpha}\right)^2}\left(\sum_{n=0}^{d} \frac{a^{(n)}(\tau)}{n!}(t-\tau)^n\right)\cos(\theta(t)) \qquad (6.1.10)$$

obtained by replacing the function a with the first $d+1$ terms of its Taylor series about τ. In what follows, we will use the expression $\approx$ to articulate an informal approximation analysis. It is clear that $\bar{f}_\tau \in \chi^{\tau,\theta_e}$ and, since $\frac{\alpha}{\dot{\theta}_0(\tau)}$ is small, that $\langle \chi^{\tau,\theta_e}_{n,j}, f_\tau - \bar{f}_\tau \rangle_{L^2} \approx 0, \forall (n,j)$ and therefore $P_{\chi^{\tau,\theta_e}} f_\tau \approx \bar{f}_\tau$, and therefore (6.1.6) implies that

$$\sum_{j'} Z_{0,j'}(\tau,\theta_e,f)\chi^{\tau,\theta_e}_{0,j'}(t) \approx \bar{f}_\tau(t), \quad t \in [-1,1], \tag{6.1.11}$$

which by (6.1.10) implies that

$$\sum_{j'} Z_{0,j'}(\tau,\theta_e,f)\chi^{\tau,\theta_e}_{0,j'}(t) \approx e^{-\left(\frac{\dot{\theta}_0(\tau)(t-\tau)}{\alpha}\right)^2} a(\tau)\cos(\theta(t)), \quad t \approx \tau, \tag{6.1.12}$$

which implies that

$$Z_{0,c}(\tau,\theta_e,f)\cos(\theta_e(t)) + Z_{0,s}(\tau,\theta_e,f)\sin(\theta_e(t)) \approx a(\tau)\cos(\theta(t)), \quad t \approx \tau. \tag{6.1.13}$$

Setting $\theta_\delta := \theta - \theta_e$ as the approximation error, using the cosine summation formula, we obtain

$$Z_{0,c}(\tau,\theta_e,f)\cos(\theta_e(t)) + Z_{0,s}(\tau,\theta_e,f)\sin(\theta_e(t))$$
$$\approx a(\tau)\big(\cos(\theta_\delta(t))\cos(\theta_e(t)) - \sin(\theta_\delta(t))\sin(\theta_e(t))\big).$$

However, $t \approx \tau$ implies that $\theta_\delta(t) \approx \theta_\delta(\tau)$, so that we obtain

$$Z_{0,c}(\tau,\theta_e,f)\cos(\theta_e(t)) + Z_{0,s}(\tau,\theta_e,f)\sin(\theta_e(t))$$
$$\approx a(\tau)\big(\cos(\theta_\delta(\tau))\cos(\theta_e(t)) - \sin(\theta_\delta(\tau))\sin(\theta_e(t))\big),$$

which, since $\dot{\theta}_e(t)$ positive and bounded away from 0, implies that

$$Z_{0,c}(\tau,\theta_e,f) \approx a(\tau)\cos(\theta_\delta(\tau))$$
$$Z_{0,s}(\tau,\theta_e,f) \approx -a(\tau)\sin(\theta_\delta(\tau)).$$

Consequently, writing

$$a(\tau,\theta_e,f) := \sqrt{Z^2_{0,c}(\tau,\theta_e,f) + Z^2_{0,s}(\tau,\theta_e,f)}$$
$$\delta\theta(\tau,\theta_e,f) := \text{atan2}\big(-Z_{0,s}(\tau,\theta_e,f), Z_{0,c}(\tau,\theta_e,f)\big), \tag{6.1.14}$$

we obtain that $a(\tau,\theta_e,f) \approx a(\tau)$ and $\delta\theta(\tau,\theta_e,f) \approx \theta_\delta(\tau)$. We will therefore use $a(\tau,\theta_e,f)$ to estimate the amplitude $a(\tau)$ of the mode f using the estimate θ_e and $\delta\theta(\tau,\theta,f)$ to estimate the mode phase θ through $\theta(\tau) = \theta_e(\tau) + \theta_\delta(\tau) \approx \theta_e(\tau) + \delta\theta(\tau,\theta_e,f)$. Unless otherwise specified, Equation (6.1.14) will take $d = 2$. Experimental evidence indicates that $d = 2$ is a sweet spot in the sense that $d = 0$ or $d = 1$ yields less fitting power, while larger d entails less stability. Iterating this refinement process will allow us to achieve near machine-precision accuracies in our phase/amplitude estimates. See sub-figures (1) and (2) of the top and bottom of Fig. 6.7 for illustrations of $a(t)$, $a(\tau,\theta_e,v)(t)$, $\theta(t) - \theta_e(t)$ and $\delta\theta(\tau,\theta_e,v)(t)$ corresponding to the first mode v_1 of the signals v displayed in Figs. 6.2.4 and 6.3.4.

6.2   The Lowest Instantaneous Frequency

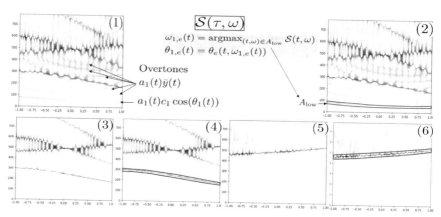

Figure 6.4:   Max-squeezing with the EKG base waveform and derivation of the instantaneous phase estimates $\theta_{i,e}$. (1,2) $(\tau, \omega) \to \mathcal{S}(\tau, \omega, v)$ and identification of A_{low}, (3, 4) $(\tau, \omega) \to \mathcal{S}(\tau, \omega, v - v_{1,e})$ and identification of its A_{low}, (5,6) $(\tau, \omega) \to \mathcal{S}(\tau, \omega, v - v_{1,e} - v_{2,e})$ and identification of its A_{low}.

We will use the max-pool network illustrated in the right-hand side of Fig. 5.3 and the module of Sect. 6.1 to design a module taking a signal v as input and producing, as output, an estimate of the instantaneous phase $\theta_{\text{low}}(v)$ of the mode of v having the lowest instantaneous frequency. We restrict our presentation to the situation where the instantaneous frequencies $\dot{\theta}_i$ do not cross each other. The main steps of the computation performed by this module are as follows. Let $\mathcal{S}(\tau, \omega, v)$ be the max-pool energy defined as in (5.4.15), where now it is useful to indicate its dependence on v.

Let A_{low} be a subset of the time–frequency domain (τ, ω) identified (as in Fig. 6.4.2) as a narrow sausage around the lowest instantaneous frequency defined by the local maxima of the $\mathcal{S}(\tau, \omega, v)$. If no modes can be detected (above a given threshold) in $\mathcal{S}(\tau, \omega, v)$, then we set $\theta_{\text{low}}(v) = \varnothing$. Otherwise we let

$$\omega_{\text{low}}(\tau) := \omega_e\left(\tau, \text{argmax}_{\omega:(\tau,\omega)\in A_{\text{low}}} \mathcal{S}(\tau, \omega)\right) \tag{6.2.1}$$

be the estimated instantaneous frequency of the mode having the lowest instantaneous frequency and, with θ_e defined as in (5.4.2), let

$$\theta_{\text{low}}(\tau) := \theta_e(\tau, \omega_{\text{low}}(\tau)) \tag{6.2.2}$$

be the corresponding estimated instantaneous phase (obtained as in (5.5.1)).

6.3 The Iterated Micro-Local KMD Algorithm

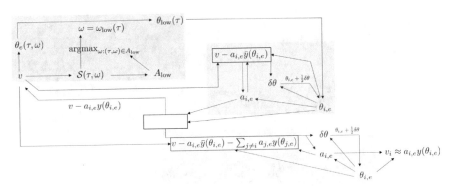

Figure 6.5: Modular representation of Algorithm 2, described in this section. The blue module represents the estimation of the lowest frequency as illustrated in Fig. 6.4. The brown module represents the iterative estimation of the mode with lowest instantaneous frequency of lines 10 through 14 of Algorithm 2. The yellow module represents the iterative refinement of all the modes in lines 21 through 27. The brown and yellow modules used to refine phase/amplitude estimates use the same code.

The method of estimating the lowest instantaneous frequency, described in Sect. 6.2, provides a foundation for the iterated micro-local KMD algorithm, Algorithm 2. We now describe Algorithm 2, presented in its modular representation in Fig. 6.5, using Figs. 6.4, 6.6 and 6.7. To that end, let

$$y(t) = c_1 \cos(t) + \sum_{n=2}^{\infty} c_n \cos(nt + d_n) \tag{6.3.1}$$

be the Fourier representation of the base waveform y (which, without loss of generality, has been shifted so that the first sine coefficient is zero) and write

$$\bar{y}(t) := y(t) - c_1 \cos(t) \tag{6.3.2}$$

for its overtones.

Let us describe how lines 1 to 19 provide refined estimates for the amplitude and the phase of each mode $v_i, i \in \{1, \ldots, m\}$ of the signal v. Although the overtones of y prevent us from simultaneously approximating all the instantaneous frequencies $\dot{\theta}_i$ from the max-pool energy of the signal v, since the lowest mode $v_{\text{low}} = a_{\text{low}} y(\theta_{\text{low}})$ can be decomposed into the sum $v_{\text{low}} = a_{\text{low}} c_1 \cos(\theta_{\text{low}}) + a_{\text{low}} \bar{y}(\theta_{\text{low}})$ of a signal $a_{\text{low}} c_1 \cos(\theta_{\text{low}})$ with a cosine

---

[1] This repeat loop, used to refine the estimates, is optional. Also, all statements in Algorithms with dummy variable τ imply a loop over all values of τ in the mesh $\mathcal{T}$.

Algorithm 2 Iterated micro-local KMD

1: $i \leftarrow 1$
2: $v^{(1)} \leftarrow v$
3: **while** true **do**
4:      **if** $\theta_{\text{low}}(v^{(i)}) = \varnothing$ **then**
5:          break loop
6:      **else**
7:          $\theta_{i,e} \leftarrow \theta_{\text{low}}(v^{(i)})$
8:      **end if**
9:      $a_{i,e}(\tau) \leftarrow 0$
10:     **repeat**
11:         **for** j in $\{1, ..., i\}$ **do**
12:             $v_{j,\text{res}} \leftarrow v - a_{j,e}\bar{y}(\theta_{j,e}) - \sum_{k \neq j, k \leqslant i} a_{k,e} y(\theta_{k,e})$
13:             $a_{j,e}(\tau) \leftarrow a(\tau, \theta_{j,e}, v_{j,\text{res}})/c_1$
14:             $\theta_{j,e}(\tau) \leftarrow \theta_{j,e}(\tau) + \frac{1}{2}\delta\theta(\tau, \theta_{j,e}, v_{j,\text{res}})$
15:         **end for**
16:     **until** $\sup_{i,\tau} \left|\delta\theta(\tau, \theta_{i,e}, v_{i,\text{res}})\right| < \epsilon_1$
17:     $v^{(i+1)} \leftarrow v - \sum_{j \leqslant i} a_{j,e} y(\theta_{j,e})$
18:     $i \leftarrow i + 1$
19: **end while**
20: $m \leftarrow i - 1$
21: **repeat**
22:     **for** i in $\{1, ..., m\}^1$ **do**
23:         $v_{i,\text{res}} \leftarrow v - a_{i,e}\bar{y}(\theta_{i,e}) - \sum_{j \neq i} a_{j,e} y(\theta_{j,e})$
24:         $a_{i,e}(\tau) \leftarrow a(\tau, \theta_{i,e}, v_{i,\text{res}})$
25:         $\theta_{i,e}(\tau) \leftarrow \theta_{i,e}(\tau) + \frac{1}{2}\delta\theta(\tau, \theta_{i,e}, v_{i,\text{res}})$
26:     **end for**
27: **until** $\sup_{j,\tau} \left|\delta\theta(\tau, \theta_{j,e}, v_{j,\text{res}})\right| < \epsilon_2$
28: Return the modes $v_{i,e}(t) \leftarrow a_{i,e}(t)y(\theta_{i,e}(t))$ for $i = 1, ..., m$

---

waveform plus the signal $a_{\text{low}}\bar{y}(\theta_{\text{low}})$ containing its higher frequency overtones, the method of Sect. 6.2 can be applied to obtain an estimate $\theta_{\text{low},e}$ of θ_{low} and (6.1.14) can be applied to obtain an estimate $a_{\text{low},e}c_1$ of $a_{\text{low}}c_1$ producing an estimate $a_{\text{low},e}c_1 \cos(\theta_{\text{low},e})$ of the primary component $a_{\text{low}}c_1 \cos(\theta_{\text{low}})$ of the first mode. Since c_1 is known, this estimate produces the estimate $a_{\text{low},e}\bar{y}(\theta_{\text{low},e})$ for the overtones of the lowest mode. Recall that we calculate all quantities over the interval $[-1, 1]$ in this setting. Estimates near the borders, -1 and 1, will be less precise but will be refined in the following loops. To improve the accuracy of this estimate, in lines 13 and 14 the micro-local KMD of Sect. 6.1 is iteratively applied to the residual signal of every previously identified mode $v_{j,\text{res}} \leftarrow v - a_{j,e}\bar{y}(\theta_{j,e}) - \sum_{k \neq j, k \leqslant i} a_{k,e} y(\theta_{k,e})$, consisting of the signal v with the estimated modes $k \neq j$ as well as the overtones of estimated mode j removed. This residual is the sum of the estimation of the isolated base frequency component of v_j and $\sum_{j>i} v_j$. The rate parameter $1/2$ in line 14 is to avoid overcorrecting the phase estimates, while the parameters ϵ_1 and ϵ_2 in lines 16 and 27 are pre-specified accuracy thresholds. The resulting estimated lower modes are then removed from the signal to determine the residual $v^{(i+1)} := v - \sum_{j \leqslant i} a_{j,e} y(\theta_{j,e})$ in line 17.

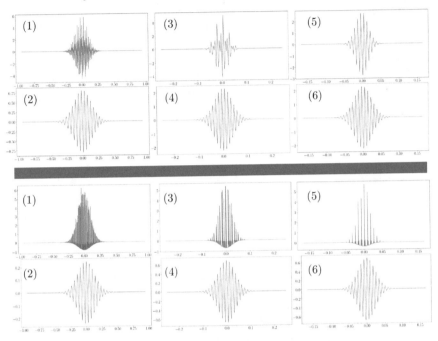

Figure 6.6: Top: v is as in Fig. 6.2 (the base waveform is triangular). Bottom: v is as in Fig. 6.3 (the base waveform is EKG). Both top and bottom: $d = 2$, (1) The windowed signal v_τ, (2) $\lim_{\sigma \downarrow 0} \mathbb{E}\left[\xi_{\tau,\theta_{1,e}} \mid \xi_{\tau,\theta_{1,e}} + \xi_\sigma = v_\tau\right]$, (3) $(v - v_{1,e})_\tau$, (4) $\lim_{\sigma \downarrow 0} \mathbb{E}\left[\xi_{\tau,\theta_{2,e}} \mid \xi_{\tau,\theta_{2,e}} + \xi_\sigma = (v - v_{1,e})_\tau\right]$, (5) $(v - v_{1,e} - v_{2,e})_\tau$, (6) $\lim_{\sigma \downarrow 0} \mathbb{E}\left[\xi_{\tau,\theta_{3,e}} \mid \xi_{\tau,\theta_{3,e}} + \xi_\sigma = (v - v_{1,e} - v_{2,e})_\tau\right]$.

Iterating this process, we peel off an estimate $a_{i,e} y(\theta_{i,e})$ of the mode corresponding to the lowest instantaneous frequency of the residual $v^{(i)} := v - \sum_{j \leqslant i-1} a_{j,e} y(\theta_{j,e})$ of the signal v obtained in line 17, removing the interference of the first $i-1$ modes, including their overtones, in our estimate of the instantaneous frequency and phase of the i-th mode. See Fig. 6.4 for the evolution of the A_{low} sausage as these modes are peeled off. See sub-figures (3) and (5) of the top and bottom of Fig. 6.6 for the results of peeling off the first two estimated modes of the signal v corresponding to both Figs. 6.2 and 6.3 and sub-figures (4) and (6) for the results of the corresponding projections in (6.1.5). See sub-figures (3) and (4) of the top and bottom of Fig. 6.7 for amplitude and its estimate of the results of peeling off the first estimated mode

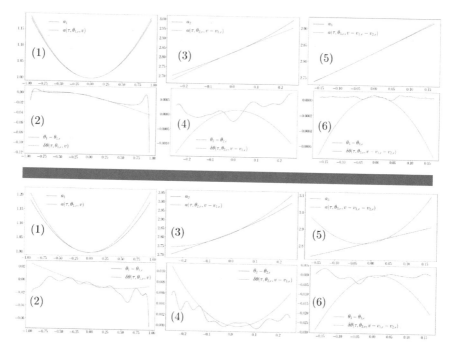

Figure 6.7: Top: v is as in Fig. 6.2 (the base waveform is triangular). Bottom: v is as in Fig. 6.3 (the base waveform is EKG). Both top and bottom: $\tau = 0$. (1) the amplitude of the first mode $a_1(t)$ and its local Gaussian regression estimation $a(\tau, \theta_{1,e}, v)(t)$, (2) the error in estimated phase of the first mode $\theta_1(t) - \theta_{1,e}(t)$ and its local Gaussian regression $\delta\theta(\tau, \theta_{1,e}, v)(t)$, (3, 4) are as (1,2) with v and $\theta_{1,e}$ replaced by $v - v_{1,e}$ and $\theta_{2,e}$, (5,6) are as (1,2) with v and $\theta_{1,e}$ replaced by $v - v_{1,e} - v_{2,e}$ and $\theta_{3,e}$.

and sub-figures (5) and (6) corresponding to peeling off the first two estimated modes of the signal v corresponding to both Figs. 6.2 and 6.3.

After the amplitude/phase estimates $a_{i,e}, \theta_{i,e}, i \in \{1, \ldots, m\}$ have been obtained in lines 1 to 19, we have the option to further improve our estimates in a final optimization loop in lines 21 to 27. This option enables us to achieve even higher accuracies by iterating the micro-local KMD of Sect. 6.1 on the residual signals $v_{i,\text{res}} \leftarrow v - a_{i,e}\bar{y}(\theta_{i,e}) - \sum_{j \neq i} a_{j,e}y(\theta_{j,e})$, consisting of the signal v with all the estimated modes $j \neq i$ and estimated overtones of the mode i removed.

The proposed algorithm can be further improved by (1) applying a Savitsky-Golay filter to locally smooth (de-noise) the curves corresponding to each estimate $\theta_{i,e}$ (which corresponds to refining our phase estimates through GPR filtering) (2) starting with a larger α (to decrease interference from other modes/overtones) and slowly reducing its value in the optional final refinement loop (to further localize our estimates after other components, and hence interference, have been mostly eliminated).

6.4 Numerical Experiments

Here we present results for both the triangle and EKG base waveform examples. As discussed in the previous section, these results are visually displayed in Figs. 6.6 and 6.7.

6.4.1 Triangle Wave Example

The base waveform is the triangle wave displayed in Fig. 6.1. We observe the signal v on a mesh spanning $[-1, 1]$ spaced at intervals of $\frac{1}{5000}$ and aim to recover each mode v_i over this time mesh. We take $\alpha = 25$ within the first refinement loop corresponding to lines 1 to 19 and slowly decreased it to 6 in the final loop corresponding to lines 21 to 27. The amplitudes and frequencies of each of the modes are shown in Fig. 6.2. The recovery errors of each mode as well as their amplitude and phase functions over the whole interval $[-1, 1]$ and the interior third $[-\frac{1}{3}, \frac{1}{3}]$ are displayed in Tables 6.1 and 6.2, respectively. In the interior third of the interval, errors were found to be on the order of 10^{-9} for the first signal component and approximately 10^{-7} for the higher two. However, over the full interval, the corresponding figures are in the 10^{-4} and 10^{-3} ranges due to recovery errors near the boundaries, -1 and 1, of the interval. Still, a plot superimposing v_i and $v_{i,e}$ would visually appear to be one curve over $[-1, 1]$ due to the negligible recovery errors.

Mode	$\dfrac{\|v_{i,e}-v_i\|_{L^2}}{\|v_i\|_{L^2}}$	$\dfrac{\|v_{i,e}-v_i\|_{L^\infty}}{\|v_i\|_{L^\infty}}$	$\dfrac{\|a_{i,e}-a_i\|_{L^2}}{\|a_i\|_{L^2}}$	$\|\theta_{i,e}-\theta_i\|_{L^2}$
$i = 1$	5.47×10^{-4}	3.85×10^{-3}	2.80×10^{-4}	4.14×10^{-5}
$i = 2$	6.42×10^{-4}	2.58×10^{-3}	3.80×10^{-5}	1.85×10^{-4}
$i = 3$	5.83×10^{-4}	6.29×10^{-3}	2.19×10^{-4}	6.30×10^{-5}

Table 6.1: Signal component recovery errors in the triangle base waveform example over $[-1, 1]$.

Mode	$\dfrac{\|v_{i,e}-v_i\|_{L^2}}{\|v_i\|_{L^2}}$	$\dfrac{\|v_{i,e}-v_i\|_{L^\infty}}{\|v_i\|_{L^\infty}}$	$\dfrac{\|a_{i,e}-a_i\|_{L^2}}{\|a_i\|_{L^2}}$	$\|\theta_{i,e}-\theta_i\|_{L^2}$
$i = 1$	1.00×10^{-8}	2.40×10^{-8}	7.08×10^{-9}	6.52×10^{-9}
$i = 2$	2.74×10^{-7}	2.55×10^{-7}	1.87×10^{-8}	2.43×10^{-7}
$i = 3$	2.37×10^{-7}	3.67×10^{-7}	1.48×10^{-7}	1.48×10^{-7}

Table 6.2: Signal component recovery errors in the triangle base waveform example over $[-\frac{1}{3}, \frac{1}{3}]$.

6.4.2 EKG Wave Example

The base waveform is the EKG wave displayed in Fig. 6.1. We use the same discrete mesh as in the triangle case. Here, we took $\alpha = 25$ in the loop corresponding to lines 1 to 19 and slowly decreased it to 15 in the final loop corresponding to lines 21 to 27. The amplitudes and frequencies of each of the modes are shown in Fig. 6.3, while the recovery error of each mode, as well as their amplitude and phase functions, are shown both over the whole interval $[-1, 1]$ and the interior third $[-\frac{1}{3}, \frac{1}{3}]$ in Tables 6.3 and 6.4, respectively. Within the interior third of the interval, amplitude and phase relative errors are found to be on the order of 10^{-4} to 10^{-5} in this setting. However, over $[-1, 1]$, the mean errors are more substantial, with amplitude and phase estimates in the 10^{-1} to 10^{-3} range. Note the high error rates in L^{∞} stemming from errors in placement of the tallest peak (the region around which is known as the R wave in the EKG community). In the center third of the interval, $v_{i,e}$ and v_i are visually indistinguishable due to the small recovery errors.

Mode	$\dfrac{\|v_{i,e}-v_i\|_{L^2}}{\|v_i\|_{L^2}}$	$\dfrac{\|v_{i,e}-v_i\|_{L^\infty}}{\|v_i\|_{L^\infty}}$	$\dfrac{\|a_{i,e}-a_i\|_{L^2}}{\|a_i\|_{L^2}}$	$\|\theta_{i,e}-\theta_i\|_{L^2}$
$i=1$	5.66×10^{-2}	1.45×10^{-1}	4.96×10^{-3}	8.43×10^{-3}
$i=2$	4.61×10^{-2}	2.39×10^{-1}	2.35×10^{-2}	1.15×10^{-2}
$i=3$	1.34×10^{-1}	9.39×10^{-1}	9.31×10^{-3}	2.69×10^{-2}

Table 6.3: Signal component recovery errors on $[-1, 1]$ in the EKG base waveform example.

Mode	$\dfrac{\|v_{i,e}-v_i\|_{L^2}}{\|v_i\|_{L^2}}$	$\dfrac{\|v_{i,e}-v_i\|_{L^\infty}}{\|v_i\|_{L^\infty}}$	$\dfrac{\|a_{i,e}-a_i\|_{L^2}}{\|a_i\|_{L^2}}$	$\|\theta_{i,e}-\theta_i\|_{L^2}$
$i=1$	1.80×10^{-4}	3.32×10^{-4}	3.52×10^{-5}	2.85×10^{-5}
$i=2$	4.35×10^{-4}	5.09×10^{-4}	3.35×10^{-5}	7.18×10^{-5}
$i=3$	3.63×10^{-4}	1.08×10^{-3}	7.23×10^{-5}	6.26×10^{-5}

Table 6.4: Signal component recovery errors on $[-\frac{1}{3}, \frac{1}{3}]$ in the EKG base waveform example.

Chapter 7

Unknown Base Waveforms

In this chapter we consider the extension, Problem 2, of the mode recovery problem, Problem 1, to the case where the periodic base waveform of each mode is unknown and may be different across modes. That is, given the observation

$$v(t) = \sum_{i=1}^{m} a_i(t) y_i\big(\theta_i(t)\big), \quad t \in [-1, 1], \tag{7.0.1}$$

recover the modes $v_i := a_i(t) y_i\big(\theta_i(t)\big)$. To avoid ambiguities caused by overtones when the waveforms y_i are not only non-trigonometric but also unknown, we will assume that the corresponding functions $(k\dot{\theta}_i)_{t\in[-1,1]}$ and $(k'\dot{\theta}_{i'})_{t\in[-1,1]}$ are distinct for $i \neq i'$ and $k, k' \in \mathbb{N}^*$, that is, they may be equal for some t but not for all t. We represent the i-th base waveform y_i through its Fourier series

$$y_i(t) = \cos(t) + \sum_{k=2}^{k_{\max}} \big(c_{i,(k,c)} \cos(kt) + c_{i,(k,s)} \sin(kt)\big), \tag{7.0.2}$$

that, without loss of generality has been scaled and translated. Moreover, since we operate in a discrete setting, without loss of generality we can also truncate the series at a finite level $k_{\max}$, which is naturally bounded by the inverse of the resolution of the discretization in time. To illustrate our approach, we consider the signal $v = v_1 + v_1 + v_3$ and its corresponding modes $v_i := a_i(t) y_i\big(\theta_i(t)\big)$ displayed in Fig. 7.1, where the corresponding base waveforms y_1, y_2 and y_3 are shown in Fig. 7.2 and described in Sect. 7.1.

7.0.1 Micro-Local Waveform KMD

We now describe the micro-local *waveform* KMD, Algorithm 3, which takes as inputs a time τ, estimated instantaneous amplitude and phase functions

© The Author(s), under exclusive license to Springer Nature Switzerland AG 2021
H. Owhadi et al., *Kernel Mode Decomposition and the Programming of Kernels*, Surveys and Tutorials in the Applied Mathematical Sciences 8, https://doi.org/10.1007/978-3-030-82171-5_7

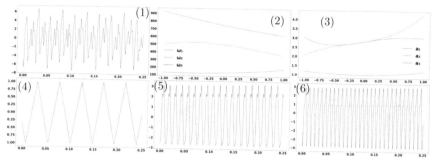

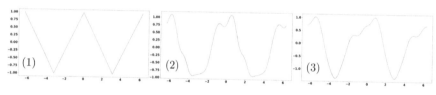

Figure 7.1: (1) Signal v (the signal is defined over $[-1, 1]$ but displayed over $[0, 0.4]$ for visibility), (2) Instantaneous frequencies $\omega_i := \dot{\theta}_i$, (3) Amplitudes a_i, (4, 5, 6) Modes v_1, v_2, v_3 over $[0, 0.4]$ (mode plots have also been zoomed in for visibility).

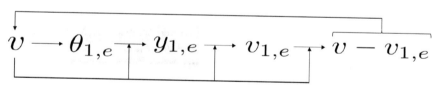

Figure 7.2: (1) y_1, (2) y_2, (3) y_3.

$$v \longrightarrow \theta_{1,e} \longrightarrow y_{1,e} \longrightarrow v_{1,e} \longrightarrow v - v_{1,e}$$

Figure 7.3: High level structure of Algorithm 3 for the case when the waveforms are unknown.

$t \to a(t), \theta(t)$, and a signal v, and outputs an estimate of the waveform $y(t)$ associated with the phase function θ. The proposed approach is a direct extension of the one presented in Sect. 6.1 and the shaded part of Fig. 7.3 shows the new block which will be added to Algorithm 2, the algorithm designed for the case when waveforms are non-trigonometric and known. As described below this new block produces an estimator $y_{i,e}$ of the waveform y_i from an estimate $\theta_{i,e}$ of the phase θ_i.

Given $\alpha > 0$, $\tau \in [-1, 1]$, and differentiable function $t \to \theta(t)$, define the Gaussian process

$$\xi^y_{\tau,\theta}(t) = e^{-\left(\frac{\dot{\theta}_0(\tau)(t-\tau)}{\alpha}\right)^2}\left(X^y_{1,c}\cos\left(\theta(t)\right) + \sum_{k=2}^{k_{\max}}\left(X^y_{k,c}\cos\left(k\theta(t)\right) + X^y_{k,s}\sin\left(k\theta(t)\right)\right)\right),$$

(7.0.3)

where $X^y_{1,c}$, $X^y_{k,c}$, and $X^y_{k,s}$ are independent $\mathcal{N}(0,1)$ random variables. Let

$$v_\tau(t) := e^{-\left(\frac{\dot{\theta}_0(\tau)(t-\tau)}{\alpha}\right)^2}v(t), \quad \tau \in [-1, 1],$$

(7.0.4)

be the windowed signal, and define

$$Z^y_{k,j}(\tau, \theta, v) := \lim_{\sigma \downarrow 0} \mathbb{E}\left[X^y_{k,j}\big|\xi^y_{\tau,\theta} + \xi_\sigma = v_\tau\right],$$

(7.0.5)

and, for $k \in \{2, \dots, k_{\max}\}$, $j \in \{c, s\}$, let

$$c_{k,j}(\tau, \theta, v) := \frac{Z^y_{k,j}(\tau, \theta, v)}{Z^y_{1,c}(\tau, \theta, v)}.$$

(7.0.6)

When the assumed phase function $\theta := \theta_{i,e}$ is close to the phase function θ_i of the i-th mode of the signal v in the expansion (7.0.1), $c_{k,j}(\tau, \theta_{i,e}, v)$ yields an estimate of the Fourier coefficient $c_{i,(k,j)}$ (7.0.2) of the i-th base waveform y_i at time $t = \tau$. This waveform recovery is susceptible to error when there is interference in the overtone frequencies (that is for the values of τ at which $j_1\dot{\theta}_{i_1} \approx j_2\dot{\theta}_{i_2}$ for $i_1 < i_2$). However, since the coefficient $c_{i,(k,j)}$ is independent of time, we can overcome this by computing $c_{k,j}(\tau, \theta_{i,e}, v)$ at each time τ and take the most common approximate value over all τ as follows. Let $T \subset [-1, 1]$ be the finite set of values of τ used in the numerical discretization of the time axis with $N := |T|$ elements. For an interval $I \subset \mathbb{R}$, let

$$T_I := \{\tau \in T \mid c_{k,j}(\tau, \theta_{i,e}, v) \in I\},$$

(7.0.7)

and let $N_I := |T_I|$ denote the number of elements of T_I. Let $I_{\max}$ be a maximizer of the function $I \to N_I$ over intervals of fixed width L, and define the estimate

$$c_{k,j}(\theta_{i,e}, v) := \begin{cases} \frac{1}{N_{I_{\max}}}\sum_{\tau \in T_{I_{\max}}} c_{k,j}(\tau, \theta_{i,e}, v) & , \quad \frac{N_{I_{\max}}}{N} \geq 0.05 \\ 0 & , \quad \frac{N_{I_{\max}}}{N} < 0.05 \end{cases},$$

(7.0.8)

of the Fourier coefficient $c_{i,(k,j)}$ to be the average of the values of $c_{k,j}(\tau, \theta_{i,e}, v)$ over $\tau \in T_{I_{\max}}$. The interpretation of the selection of the cutoff 0.05 is as follows: if $\frac{N_{I_{\max}}}{N}$ is small, then there is interference in the overtones at all time $[-1, 1]$ and no information may be obtained about the corresponding Fourier coefficient. When the assumed phase function is near that of the lowest frequency mode v_1, which we write $\theta := \theta_{1,e}$, Fig. 7.4.2 and 4 shows zoomed-in histograms of the functions $\tau \to c_{(3,c)}(\tau, \theta_{1,e}, v)$ and $\tau \to c_{(3,s)}(\tau, \theta_{1,e}, v)$ displayed in Fig. 7.4.1 and 3.

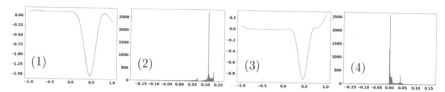

Figure 7.4: (1) A plot of the function $\tau \to c_{(3,c)}(\tau, \theta_{1,e}, v)$, (2) A histogram (cropping outliers) with bin width 0.002 of $c_{(3,c)}(\tau, \theta_{1,e}, v)$ values. The true value $c_{1,(3,c)}$ is $1/9$ since y_1 is a triangle wave. (3) A plot of the function $\tau \to c_{(3,s)}(\tau, \theta_{1,e}, v)$ (2) A histogram (cropping outliers) with bin width 0.002 of $c_{(3,s)}(\tau, \theta_{1,e}, v)$ values. The true value $c_{1,(3,s)}$ of this overtone is 0.

On the Interval Width L   In our numerical experiments, the recovered modes and waveforms show little sensitivity to the choice of L. In particular, we set L to be 0.002, whereas widths between 0.001 and 0.01 yield similar results. The rationale for the rough selection of the value of L is as follows. Suppose $v = \cos(\omega t)$ and $v' = v + \cos(1.5\omega t)$. Define the quantity

$$\max_{\tau}\left(c_{2,c}(\tau, \theta, v') - c_{2,c}(\tau, \theta, v)\right), \qquad (7.0.9)$$

with the intuition of approximating the maximum corruption by the $\cos(1.5\omega t)$ term in the estimated first overtone. This quantity provides a good choice for L and is mainly dependent on the selection of α and marginally on ω. For our selection of $\alpha = 10$, we numerically found its value to be approximately 0.002.

7.0.2   Iterated Micro-Local KMD with Unknown Waveforms Algorithm

Except for the steps discussed in Sect. 7.0.1, Algorithm 3 is identical to Algorithm 2. As illustrated in Fig. 7.3, we first identify the lowest frequency of the cosine component of each mode (lines 6 and 7 in Algorithm 3). Next, from lines 10 to 18, using Eq. 6.3.2 we execute a similar refinement loop as in Algorithm 2 with the addition of an application of micro-local waveform KMD on lines 15 and 16 to estimate base waveforms. Finally, once each mode has been identified, we again apply waveform estimation in lines 28–29 (after nearly eliminating other modes and reducing interference in overtones for higher accuracies).

7.1   Numerical Experiments

To illustrate this learning of the base waveform of each mode, we take $v(t) = \sum_{i=1}^{3} a_i(t) y_i(\theta_i(t))$, where the lowest frequency mode $a_1(t) y_1(\theta_1(t))$ has the (unknown) triangle waveform y_1 of Fig. 6.1. We determine the waveforms $y_i, i = 2, 3$, randomly by setting $c_{i,(k,j)}$ to be zero with probability $1/2$ or to be

---

Algorithm 3 Iterated micro-local KMD with unknown waveforms

1: $i \leftarrow 1$ and $v^{(1)} \leftarrow v$
2: **while** true **do**
3:    **if** $\theta_{\text{low}}(v^{(i)}) = \varnothing$ **then**
4:      break loop
5:    **else**
6:      $\theta_{i,e} \leftarrow \theta_{\text{low}}(v^{(i)})$
7:      $y_{i,e} \leftarrow \cos(t)$
8:    **end if**
9:    $a_{i,e}(\tau) \leftarrow 0$
10:   **repeat**
11:      **for** l in $\{1,...,i\}$ **do**
12:        $v_{l,\text{res}} \leftarrow v - a_{l,e}\bar{y}_{l,e}(\theta_{l,e}) - \sum_{k \neq l, k \leqslant i} a_{k,e} y_{l,e}(\theta_{k,e})$
13:        $a_{l,e}(\tau) \leftarrow a(\tau, \theta_{l,e}, v_{l,\text{res}})/c_1$
14:        $\theta_{l,e}(\tau) \leftarrow \theta_{l,e}(\tau) + \frac{1}{2}\delta\theta(\tau, \theta_{l,e}, v_{l,\text{res}})$
15:        $c_{l,(k,j),e} \leftarrow c_{k,j}(\theta_{l,e}, v_{l,\text{res}})$
16:        $y_{l,e}(\cdot) \leftarrow \cos(\cdot) + \sum_{k=2}^{k_{\max}}(c_{l,(k,c),e}\cos(k\cdot) + c_{l,(k,s),e}\sin(k\cdot))$
17:      **end for**
18:    **until** $\sup_{l,\tau}|\delta\theta(\tau, \theta_{l,e}, v_{l,\text{res}})| < \epsilon_1$
19:    $v^{(i+1)} \leftarrow v - \sum_{j \leqslant i} a_{j,e} y_{i,e}(\theta_{j,e})$
20:    $i \leftarrow i + 1$
21: **end while**
22: $m \leftarrow i - 1$
23: **repeat**
24:    **for** i in $\{1, \ldots, m\}^1$ **do**
25:      $v_{i,\text{res}} \leftarrow v - a_{i,e}\bar{y}_{i,e}(\theta_{i,e}) - \sum_{j \neq i} a_{j,e} y_{j,e}(\theta_{j,e})$
26:      $a_{i,e}(\tau) \leftarrow a(\tau, \theta_{i,e}, v_{i,\text{res}})$
27:      $\theta_{i,e}(\tau) \leftarrow \theta_{i,e}(\tau) + \frac{1}{2}\delta\theta(\tau, \theta_{i,e}, v_{i,\text{res}})$
28:      $c_{i,(k,j),e} \leftarrow c_{k,j}(\theta_{i,e}, v - \sum_{j \neq i} a_{j,e} y_{j,e}(\theta_{j,e}))$
29:      $y_{i,e}(\cdot) \leftarrow \cos(\cdot) + \sum_{k=2}^{k_{\max}}(c_{i,(k,c),e}\cos(k\cdot) + c_{i,(k,s),e}\sin(k\cdot))$
30:    **end for**
31: **until** $\sup_{i,\tau}|\delta\theta(\tau, \theta_{i,e}, v_{i,\text{res}})| < \epsilon_2$
32: Return the modes $v_{i,e}(t) \leftarrow a_{i,e}(t)y(\theta_{i,e}(t))$ for $i = 1, ..., m$

---

a random sample from $\mathcal{N}(0, 1/k^4)$ with probability $1/2$, for $k \in \{2, \ldots, 7\}$ and $j \in \{c, s\}$. The waveforms y_1, y_2, y_3 thus obtained are illustrated in Fig. 7.2. The modes v_1, v_2, v_3, their amplitudes and instantaneous frequencies are shown in Fig. 7.1.

We use the same mesh and the same value of α values as in Sect. 6.4.1. The main source of error for the recovery of the first mode's base waveform stems from the fact that a triangle wave has an infinite number of overtones,

---

[1] This repeat loop, used to refine the estimates, is optional. Also, all statements in Algorithms with dummy variable τ imply a loop over all values of τ in the mesh $\mathcal{T}$.

Mode	$\dfrac{\|v_{i,e}-v_i\|_{L^2}}{\|v_i\|_{L^2}}$	$\dfrac{\|v_{i,e}-v_i\|_{L^\infty}}{\|v_i\|_{L^\infty}}$	$\dfrac{\|a_{i,e}-a_i\|_{L^2}}{\|a_i\|_{L^2}}$	$\|\theta_{i,e}-\theta_i\|_{L^2}$	$\dfrac{\|y_{i,e}-y_i\|_{L^2}}{\|y_i\|_{L^2}}$
$i = 1$	6.31×10^{-3}	2.39×10^{-2}	9.69×10^{-5}	1.41×10^{-5}	6.32×10^{-3}
$i = 2$	3.83×10^{-4}	1.08×10^{-3}	5.75×10^{-5}	1.16×10^{-4}	3.76×10^{-4}
$i = 3$	3.94×10^{-4}	1.46×10^{-3}	9.53×10^{-5}	6.77×10^{-5}	3.80×10^{-4}

Table 7.1: Signal component recovery errors over $[-1, 1]$ when the base waveforms are unknown.

while in our implementation, we estimate only the first 15 overtones. Indeed, the L^2 recovery error of approximating the first 16 tones of the triangle wave is 3.57×10^{-4}, while the full recovery errors are presented in Table 7.1. We omitted the plots of the $y_{i,e}$ as they are visually indistinguishable from those of the y_i. Note that errors are only slightly improved away from the borders as the majority of it is accounted for by the waveform recovery error.

Chapter 8

Crossing Frequencies, Vanishing Modes, and Noise

In this chapter, we address the following generalization of the mode recovery Problem 4, allowing for crossing frequencies, vanishing modes and noise. The purpose of the δ, ϵ-condition in Problem 5 is to prevent a long overlap of the instantaneous frequencies of distinct modes.

Problem 5. *For* $m \in \mathbb{N}^*$, *let* $a_1, \ldots, a_m$ *be piecewise smooth functions on* $[-1, 1]$, *and let* $\theta_1, \ldots, \theta_m$ *be strictly increasing functions on* $[-1, 1]$ *such that, for* $\epsilon > 0$ *and* $\delta \in [0, 1)$, *the length of* t *with* $\dot{\theta}_i(t)/\dot{\theta}_j(t) \in [1 - \epsilon, 1 + \epsilon]$ *is less than* δ. *Assume that* m *and the* a_i, θ_i *are unknown, and the square-integrable* 2π-*periodic base waveform* y *is known. Given the observation* $v(t) = \sum_{i=1}^m a_i(t) y(\theta_i(t)) + v_\sigma(t)$ *(for* $t \in [-1, 1]$*), where* v_σ *is a realization of white noise with variance* σ^2, *recover the modes* $v_i(t) := a_i(t) y(\theta_i(t))$.

8.1 Illustrative Examples

We will use the following two examples to illustrate our algorithm, in particular the identification of the lowest frequency $\omega_{\text{low}}(\tau)$, at each time τ, and the process of obtaining estimates of modes.

Example 8.1.1. *Consider the problem of recovering the modes of the signal* $v = v_1 + v_2 + v_3 + v_\sigma$ *shown in Fig. 8.1. Each mode has a triangular base waveform. In this example* v_3 *has the highest frequency and its amplitude vanishes over* $t > -0.25$. *The frequencies of* v_1 *and* v_2 *cross around* $t = 0.25$. $v_\sigma \sim \mathcal{N}(0, \sigma^2 \delta(t - s))$ *is white noise with standard deviation* $\sigma = 0.5$. *While the signal-to-noise ratio is* $\text{Var}(v_1 + v_2 + v_3)/\text{Var}(v_\sigma) = 13.1$, *the SNR ratio against each of the modes* $\text{Var}(v_i)/\text{Var}(v_\sigma)$, $i = 1, 2, 3$, *is 2.7, 7.7, and 10.7, respectively.*

© The Author(s), under exclusive license to Springer Nature Switzerland AG 2021
H. Owhadi et al., *Kernel Mode Decomposition and the Programming of Kernels*, Surveys and Tutorials in the Applied Mathematical Sciences 8, https://doi.org/10.1007/978-3-030-82171-5_8

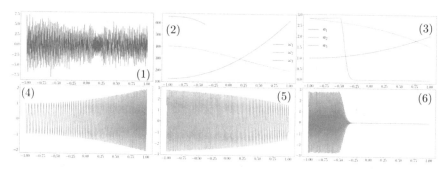

Figure 8.1: (1) Signal v, (2) Instantaneous frequencies $\omega_i := \dot{\theta}_i$, (3) Amplitudes a_i, (4, 5, 6) Modes v_1, v_2, v_3.

Example 8.1.2. *Consider the signal $v = v_1 + v_2 + v_3 + v_\sigma$ shown in Fig. 8.2. Each mode has a triangular base waveform. In this example, the vanishing mode, v_1, has the lowest frequency over $t \lesssim -0.25$ but then its amplitude vanishes over $t \gtrsim -0.25$. The frequencies of v_2 and v_3, cross around $t = 0.25$. $v_\sigma \sim \mathcal{N}(0, \sigma^2 \delta(t - s))$ is white noise with standard deviation $\sigma = 0.5$.*

Figure 8.2: (1) Signal v, (2) Instantaneous frequencies $\omega_i := \dot{\theta}_i$, (3) Amplitudes a_i.

Examples 8.1.1 and 8.1.2 of Problem 5 cannot directly be solved with Algorithm 2 (where the mode with the lowest frequency is iteratively identified and peeled off) because the lowest observed instantaneous frequency may no longer be associated with the same mode at different times in $[-1, 1]$ (due to vanishing amplitudes and crossing frequencies). Indeed, as can be seen in Fig. 8.1.2, the mode v_1 will have lowest instantaneous frequency at times prior to the intersection, i.e., over $t \lesssim 0.25$, while the lowest frequency is associated with v_2 over $t \gtrsim 0.25$. Further, in Example 8.1.2 which has modes with frequencies illustrated in Fig. 8.2.2, Fig. 8.2.3 shows that the amplitude of the mode v_1 vanishes for $t \gtrsim -0.5$ and therefore will not contribute to a lowest frequency estimation in that interval. Figure 8.2.2 implies that v_1 will appear to have the lowest instantaneous frequency for $t \lesssim -0.5$, v_2 will appear to for $t \gtrsim 0.25$, and v_3 otherwise.

The algorithms introduced in this section will address these challenges by first estimating the lowest frequency mode at each point of time in $[-1, 1]$ and dividing the domain into intervals with continuous instantaneous frequency and $\dot{\theta}_{\text{low}} \approx \omega_{\text{low}}$ in Algorithm 4. Divisions to $[-1, 1]$ can be caused by either a mode vanishing or a frequency intersection. The portions of modes corresponding to these resulting intervals with identified instantaneous frequencies are called *mode fragments*. Next, Algorithm 5 extends the domain of these fragments to the maximal domain such that the instantaneous frequency is continuous and $\dot{\theta}_{\text{low}} \approx \omega_{\text{low}}$, thus determining what are called *mode segments*. The difference between fragments and segments is elaborated in the discussion of Fig. 8.3. Furthermore, in Algorithm 6, the segments that are judged to be an artifact of noise or a mode intersection are removed. After segments are grouped by the judgment of the user of the algorithm into which belong to the same mode, they are then joined via interpolation to create estimates of full modes. Finally, in Algorithm 7, mode estimates are refined as in the final refinement loop in Algorithm 2. Python source codes are available at https://github.com/kernel-enthusiasts/Kernel-Mode-Decomposition-1D.

8.2 Identifying Modes and Segments

Algorithm 4, which follows, presents the main module $m_{\text{mode}}(v, \mathcal{V}, \mathcal{V}_{\text{seg}})$ composing Algorithm 7. The input of this module is the original signal v, a set of (estimated) *modes* $\mathcal{V} := \{v_{i,e} : [-1, 1] \to \mathbb{R}\}$, and a set $\mathcal{V}_{\text{seg}} := \{v^{i,e} : \mathsf{T}_{i,e} \to \mathbb{R}\}$ of (estimated) *segments* $v^{i,e}$, where each mode is defined in terms of its amplitude $a_{i,e}$ and phase $\theta_{i,e}$ as $v_{i,e}(t) := a_{i,e}(t)y(\theta_{i,e}(t))$, and each segment is defined in terms of its amplitude $a^{i,e}$ and phase $\theta^{i,e}$ as the function $v^{i,e}(t) := a^{i,e}(t)y(\theta^{i,e}(t))$ on its closed interval domain $\mathsf{T}_{i,e}$. In Algorithm 4 we consider a uniform mesh $\mathcal{T} \subset [-1, 1]$ with mesh spacing δt and define a *mesh interval* $[a, b] := \{t \in \mathcal{T} : a \leqslant t \leqslant b\}$, using the same notation for a mesh interval as a regular closed interval. In particular, both the modes and segments $v_{i,e}, v^{i,e}$ contain, as data, their amplitudes $a_{i,e}, a^{i,e}$ and phase functions $\theta_{i,e}, \theta^{i,e}$, while the segments additionally contain as data their domain $\mathsf{T}_{i,e}$. Moreover, their frequencies $\omega_{i,e}, \omega^{i,e}$ can also be directly extracted since they are a function of their phase functions. The output of this module is an updated set of modes $\mathcal{V}^{\text{out}}$ and segments $\mathcal{V}_{\text{seg}}^{\text{out}}$. The first step of this module (lines 2 to 5 of Algorithm 4) is to compute, for each time $\tau \in [-1, 1]$, the residual

$$v_\tau := v - \sum_{v_{i,e} \in \mathcal{V}} v_{i,e} - \sum_{v^{i,e} \in \mathcal{V}_{\text{seg}} : \tau \in \mathsf{T}_{i,e}} v_\tau^{i,e} \tag{8.2.1}$$

of the original signal after peeling off the modes and *localized segments*, where the localized segment

$$v_\tau^{i,e}(t) := a^{i,e}(\tau)e^{-\left(\frac{\omega^{i,e}(\tau)(t-\tau)}{\alpha}\right)^2} y\big((t-\tau)\omega^{i,e}(\tau) + \theta^{i,e}(\tau)\big), \quad t \in [-1, 1], \ \tau \in \mathsf{T}_{i,e}, \tag{8.2.2}$$

defined from the amplitude, phase and frequency of segment $v^{i,e}$, is well-defined on the whole domain $[-1, 1]$ when $\tau \in \mathsf{T}_{i,e}$. Extending $v_\tau^{i,e}$ so that it is defined as the zero function for $\tau \notin \mathsf{T}_{i,e}$, (8.2.1) appears more simply as

$$v_\tau := v - \sum_{\mathcal{V}} v_{i,e} - \sum_{\mathcal{V}_{\text{seg}}} v_\tau^{i,e}. \qquad (8.2.3)$$

Note that unlike previous sections where the function θ_0, common throughout many iterations, would be determining the width parameter $\dot{\theta}_0(\tau)$ in the exponential in (8.2.2), here the latest frequency estimate $\omega^{i,e}$ is used. The peeling (8.2.3) of the modes, as well as the segments, off of the signal v is to identify other segments with higher instantaneous frequencies.

Next, in line 6 of Algorithm 4, we compute the lowest instantaneous frequency $\omega_{\text{low}}(\tau, v_\tau)$ of v_τ as in (6.2.1), where A_{low} is determined either by the user or a set of rules, e.g., we identify $\omega_{\text{low}}(\tau, v_\tau)$ as the lowest frequency local maxima of the energy $\mathcal{S}(\tau, \cdot, v_\tau)$ that is greater than a set threshold ϵ_0 (in our implementations, we set this threshold as a fixed fraction of $\max_{\tau, \omega} \mathcal{S}(\tau, \omega, v)$). If no energies are detected above this given threshold in $\mathcal{S}(\tau, \cdot, v_\tau)$ we set $\omega_{\text{low}}(\tau, v_\tau) = \varnothing$. We use the abbreviation $\omega_{\text{low}}(\tau)$ for $\omega_{\text{low}}(\tau, v_\tau)$. Figure 8.3.2 shows $\omega_{\text{low}}(\tau)$ derived from $\mathcal{S}$ (Fig. 8.3.1) in Example 8.1.2. Then, using the micro-local KMD approach of Sect. 6.1 with (the maximum polynomial degree) d set to 0, lines 8 and 9 of Algorithm 4 compute an amplitude

$$a_{\text{low}}(\tau) := a(\tau, (\cdot - \tau)\omega_{\text{low}}(\tau), v) \qquad (8.2.4)$$

and phase

$$\theta_{\text{low}}(\tau) := \delta\theta(\tau, (\cdot - \tau)\omega_{\text{low}}(\tau), v) \qquad (8.2.5)$$

at $t = \tau$, using (6.1.14) applied to the locally estimated phase function $(\cdot - \tau)\omega_{\text{low}}(\tau)$ determined by the estimated instantaneous frequency $\omega_{\text{low}}(\tau)$. The approximation (8.2.5) is justified since this estimated phase function $(\cdot - \tau)\omega_{\text{low}}(\tau)$ vanishes at $t = \tau$, so that the discussion below (6.1.14) demonstrates that the updated estimated phase $0 + \delta\theta(\tau, (\cdot - \tau)\omega_{\text{low}}(\tau), v) = \delta\theta(\tau, (\cdot - \tau)\omega_{\text{low}}(\tau), v)$ is an estimate of the instantaneous phase at $t = \tau$ and frequency $\omega = \omega_{\text{low}}(\tau)$. Then $a_{\text{low}}(\tau)y(\theta_{\text{low}}(\tau))$ is an estimate, at $t = \tau$, of the mode having the lowest frequency. If $\omega_{\text{low}}(\tau) = \varnothing$, we leave a_{low} and θ_{low} undefined.

Next, let us describe how we use the values of $(\tau, \omega_{\text{low}}(\tau))$ to determine the interval domains for segments. Writing $\mathcal{T}_{\text{cut}}$ for the set of interval domains of these segments, $\mathcal{T}_{\text{cut}}$ is initially set, in line 17, to contain the single element $\mathcal{T}$, that is, the entire time mesh $\mathcal{T}$. We split an element of $\mathcal{T}_{\text{cut}}$ whenever ω_{low} is not continuous or $\dot{\theta}_{\text{low}}$ and ω_{low} are not approximately equal, as follows. If our identified instantaneous frequency around $t = \tau$ matches a single mode, we expect neither condition to be satisfied, i.e., we expect both ω_{low} to be continuous and $\dot{\theta}_{\text{low}} \approx \omega_{\text{low}}$. In our discrete implementation (lines 18 to 24), we introduce a cut between two successive points, τ_1 and τ_2, of the time mesh $\mathcal{T}$, if

$$\left| \log\left(\frac{\omega_{\text{low}}(\tau_2)}{\omega_{\text{low}}(\tau_1)} \right) \right| > \epsilon_1 \quad \text{or} \quad \left| \log\left(\frac{(\theta_{\text{low}}(\tau_2) - \theta_{\text{low}}(\tau_1))(\tau_2 - \tau_1)^{-1}}{\omega_{\text{low}}(\tau_1)} \right) \right| > \epsilon_2,$$

$$(8.2.6)$$

Algorithm 4 Lowest frequency segment identification

1: **function** $m_{\mathrm{mode}}(v, \mathcal{V}, \mathcal{V}_{\mathrm{seg}})$

2:     **for** $v^{i,e}$ in $\mathcal{V}_{\mathrm{seg}}$ **do**

3:         $v_\tau^{i,e}(t) \leftarrow a^{i,e}(\tau) e^{-\left(\frac{\omega^{i,e}(\tau)(t-\tau)}{\alpha}\right)^2} y((t-\tau)\omega^{i,e}(\tau) + \theta^{i,e}(\tau))$

4:     **end for**

5:     $v_\tau \leftarrow v - \sum_\mathcal{V} v_{i,e} - \sum_{\mathcal{V}_{\mathrm{seg}}} v_\tau^{i,e}$

6:     Get $\omega_{\mathrm{low}}(\tau, v_\tau)$ as in (6.2.1) and abbreviate it as $\omega_{\mathrm{low}}(\tau)$

7:     **if** $\omega_{\mathrm{low}}(\tau) \neq \varnothing$

8:         $a_{\mathrm{low}}(\tau) \leftarrow a(\tau, (\cdot - \tau)\omega_{\mathrm{low}}(\tau), v_\tau)$

9:         $\theta_{\mathrm{low}}(\tau) \leftarrow \delta\theta(\tau, (\cdot - \tau)\omega_{\mathrm{low}}(\tau), v_\tau)$

10:     **end if**

11:     Set $\mathcal{T}$ to be the regular time mesh with spacing δt

12:     $\mathcal{T} \leftarrow \mathcal{T} \cap \{\tau | \omega_{\mathrm{low}}(\tau) \neq \varnothing\}$

13:     **if** $\mathcal{T} = \varnothing$ **then**

14:         $\mathcal{V}_{\mathrm{seg}} \leftarrow \varnothing$

15:         **return** $\mathcal{V}, \mathcal{V}_{\mathrm{seg}}$ and goto line 34

16:     **end if**

17:     $\mathcal{T}_{\mathrm{cut}} \leftarrow \{[\min(\mathcal{T}), \max(\mathcal{T})]\}$ (Initialize the set of mesh intervals $\mathcal{T}_{\mathrm{cut}}$)

18:     **for** successive $\tau_1, \tau_2 \ (\tau_2 - \tau_1 = \delta t)$ in $\mathcal{T}$ **do**

19:         **if** $\left| \log\left(\frac{\omega_{\mathrm{low}}(\tau_2)}{\omega_{\mathrm{low}}(\tau_1)}\right) \right| > \epsilon_1$ or $\left| \log\left(\frac{(\theta_{\mathrm{low}}(\tau_2) - \theta_{\mathrm{low}}(\tau_1))(\tau_2 - \tau_1)^{-1}}{\omega_{\mathrm{low}}(\tau_1)}\right) \right| > \epsilon_2$

    then

20:             **if** $[\tau_1, \tau_2] \subset [t_1, t_2] \in \mathcal{T}_{\mathrm{cut}}$ **then**

21:                 $\mathcal{T}_{\mathrm{cut}} \leftarrow (\mathcal{T}_{\mathrm{cut}} \smallsetminus \{[t_1, t_2]\}) \cup \{[t_1, \tau_1], [\tau_2, t_2]\}$

22:             **end if**

23:         **end if**

24:     **end for**

25:     $v_{\mathrm{low}} \leftarrow a_{\mathrm{low}} y(\theta_{\mathrm{low}})$

26:     **for** $[t_1, t_2]$ in $\mathcal{T}_{\mathrm{cut}}$ **do**

27:         $v_{\mathrm{seg},[t_1', t_2']}, t_1', t_2' \leftarrow \mathrm{MODE\_EXTEND}(v, v_{\mathrm{low}}|_{[t_1, t_2]}, \mathcal{S}(\cdot, \cdot, v_\tau))$

28:         **if** $\int_{t_1'}^{t_2'} \omega_{\mathrm{low}}(\tau) d\tau > \epsilon_3$ **then**

29:             $\mathcal{V}_{\mathrm{seg}} \leftarrow \mathcal{V}_{\mathrm{seg}} \cup \{v_{\mathrm{seg},[t_1', t_2']}\}$

30:         **end if**

31:     **end for**

32:     $\mathcal{V}^{\mathrm{out}}, \mathcal{V}_{\mathrm{seg}}^{\mathrm{out}} \leftarrow \mathrm{MODE\_PROCESS}(\mathcal{V}, \mathcal{V}_{\mathrm{seg}}, \mathcal{S}(\cdot, \cdot, v_\tau))$

33:     **return** $\mathcal{V}^{\mathrm{out}}, \mathcal{V}_{\mathrm{seg}}^{\mathrm{out}}$

34: **end function**

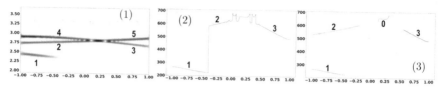

Figure 8.3: The identification of the first mode segments in Example 8.1.2 is shown. The scale of the vertical axis is $\log_{10}(\omega)$ in sub-figure (1) and ω in sub-figures (2) and (3) Segments are labeled in (1). (1) Energy $\mathcal{S}(\cdot, \cdot, v)$ (2) the identified lowest frequency at each time t with consistent segment numbering (3) identified mode segments including an artifact of the intersection, labeled as segment 0.

where ϵ_1 and ϵ_2 are pre-set thresholds. Each potential mode segment is then identified as $v_{\text{low}}|_{[t_1, t_2]}$ for some $t_1 < t_2$, $t_1, t_2 \in \mathcal{T}$.

Note that in Fig. 8.3.2, the continuous stretch of ω_{low} labeled by 2 does not correspond to the full mode segment labeled by 2 in Fig. 8.3.1, but a fragment of it. This is because the lowest frequency mode, v_1, is identified by $\omega_{\text{low}}(t)$ over $t \lesssim -0.5$. We designate this partially identified mode segment as a mode fragment. Such fragments are extended to fully identified segments (as in 2 on Fig. 8.3.3) with the MODE_EXTEND module, with pseudo-code shown in Algorithm 5. This MODE_EXTEND module iteratively extends the support, $[t_1, t_2]$, by applying, in lines 8 and 23, a max-squeezing to identify instantaneous frequencies at neighboring mesh points to the left and right of the interval $[t_1, t_2]$. The process is stopped if it is detected, in lines 10 and 25, that the extension is discontinuous in phase according to (8.2.6). This sub-module returns (maximally continuous) full mode segments. Furthermore, to remove segments that may be generated by noise or are mode intersections, in lines 26 to 31 of Algorithm 4, segments such that

$$\int_{t_1}^{t_2} \omega_{\text{low}}(\tau) d\tau \leq \epsilon_3, \qquad (8.2.7)$$

where ϵ_3 is a threshold, are removed. In our implementation, we take $\epsilon_3 := 20\pi$, corresponding to 10 full periods. Note that Fig. 8.4.2 shows those segments deemed noise at level $\epsilon_3 := 20\pi$ but which are not deemed noise at level 3π, in the step after all three modes have been estimated in Example 8.1.1. Consequently, it appears that the noise level $\epsilon_3 := 20\pi$ successfully removes most noise artifacts. Note that the mode segments in Fig. 8.4.2 are short and have quickly varying frequencies compared to those of full modes.

Next, line 32 of Algorithm 4 applies the function MODE_PROCESS, Algorithm 6, to $\mathcal{V}$ and $\mathcal{V}_{\text{seg}}$, the sets of modes and segments, as well as the energy $\mathcal{S}(\cdot, \cdot, v_\tau)$, to produce the updated sets $\mathcal{V}^{out}$ and $\mathcal{V}_{\text{seg}}^{out}$. This function utilizes a partition of a set $\mathcal{V}_{\text{group}}$, initialized to be empty, into a set of partition blocks $(\mathcal{V}_{\text{group},j})_j$, where $\mathcal{V}_{\text{group},j} \subset \mathcal{V}_{\text{group}}, \forall j$. The partition blocks consist of seg-

---

Algorithm 5 Mode fragment extension

1: **function** MODE_EXTEND$(v, v_{\text{seg}}, \mathcal{S}(\cdot, \cdot, v_\tau))$

2:   smooth $\leftarrow$ True

3:   $\tau_1 \leftarrow t_1$

4:   **while** smooth is True **do**

5:     $\theta_1 \leftarrow \theta_{\text{seg}}(\tau_1)$

6:     $\omega_1 \leftarrow \dot{\theta}_{\text{seg}}(\tau_1)$

7:     $\tau_2 \leftarrow \tau_1 - dt$

8:     $\omega_2 \leftarrow \text{argmax}_{\omega \in [(1-\varepsilon)\omega_1, (1+\varepsilon)\omega_2]} \mathcal{S}(\tau_2, \omega, v_\tau)$

9:     $\theta_2 \leftarrow \delta\theta(\tau_2, (\cdot - \tau_2)\omega_2, v_\tau)$

10:    **if** $\left| \log\left(\frac{\omega_2}{\omega_1}\right) \right| > \epsilon_1$ or $\left| \log\left(\frac{(\theta_2 - \theta_1)(\tau_2 - \tau_1)^{-1}}{\omega_1}\right) \right| > \epsilon_2$ **then**

11:       smooth $\leftarrow$ False

12:    **else**

13:      $a_2 \leftarrow a(\tau_2, (\cdot - \tau_2)\omega_2, v_\tau)$

14:      $v_{\text{seg}}(\tau_2) \leftarrow a_2 y(\theta_2)$

15:      $t_1, \tau_1 \leftarrow \tau_2$

16:    **end if**

17:   **end while**

18:   $\tau_1 \leftarrow t_2$

19:   **while** smooth is True **do**

20:     $\theta_1 \leftarrow \theta_{\text{seg}}(\tau_1)$

21:     $\omega_1 \leftarrow \dot{\theta}_{\text{seg}}(\tau_1)$

22:     $\tau_2 \leftarrow \tau_1 + dt$

23:     $\omega_2 \leftarrow \text{argmax}_{\omega \in [(1-\varepsilon)\omega_1, (1+\varepsilon)\omega_2]} \mathcal{S}(\tau_2, \omega, v_\tau)$

24:     $\theta_2 \leftarrow \delta\theta(\tau_2, (\cdot - \tau_2)\omega_2, v_\tau)$

25:    **if** $\left| \log\left(\frac{\omega_2}{\omega_1}\right) \right| > \epsilon_1$ or $\left| \log\left(\frac{(\theta_2 - \theta_1)(\tau_2 - \tau_1)^{-1}}{\omega_1}\right) \right| > \epsilon_2$ **then**

26:       smooth $\leftarrow$ False

27:    **else**

28:      $a_2 \leftarrow a(\tau_2, (\cdot - \tau_2)\omega_2, v_\tau)$

29:      $v_{\text{seg}}(\tau_2) \leftarrow a_2 y(\theta_2)$

30:      $t_2, \tau_1 \leftarrow \tau_2$

31:    **end if**

32:   **end while**

33:   **return** v_{seg}, t_1, t_2

34: **end function**

---

ments that have been identified as corresponding to the same mode, indexed locally by j. Each segment in $\mathcal{V}_{\text{seg}}$ will either be discarded or placed into a partition block. When a partition block is *complete* it will be turned into a mode in $\mathcal{V}^{\text{out}}$ by interpolating instantaneous frequencies and amplitudes in the (small) missing sections of $\mathcal{T}$ and the elements of the partition block removed from

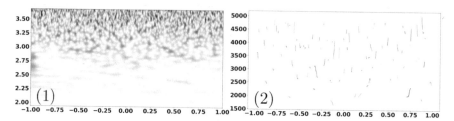

Figure 8.4: (1) Energy $\mathcal{S}(\cdot, \cdot, v - v_{1,e} - v_{2,e} - v_{3,e})$ (2) identified mode segments ($\mathcal{V}_{\text{seg}}$ obtained after the loop in Algorithm 4 on line 31).

Algorithm 6 Raw segment processing

1: **function** MODE_PROCESS($\mathcal{V}, \mathcal{V}_{\text{seg}}, \mathcal{S}(\cdot, \cdot, v_\tau)$)
2:     $\mathcal{V}_{\text{group}} \leftarrow \emptyset$
3:     **for** $v^{i,e}$ in $\mathcal{V}_{\text{seg}}$ **do**
4:         **if** $v^{i,e}$ corresponds to a mode intersection or noise **then**
5:             $\mathcal{V}_{\text{seg}} \leftarrow \mathcal{V}_{\text{seg}} \smallsetminus \{v^{i,e}\}$
6:         **else**
7:             **for** $\mathcal{V}_{\text{group},j}$ in $(\mathcal{V}_{\text{group},j'})_{j'}$ **do**
8:                 **if** $v^{i,e}$ corresponds to the same mode as $\mathcal{V}_{\text{group},j}$ **then**
9:                     $\mathcal{V}_{\text{group},j} \leftarrow \mathcal{V}_{\text{group},j} \cup \{v^{i,e}\}$
10:                     break for loop
11:                 **end if**
12:             **end for**
13:             **if** $v^{i,e}$ not added to any mode block **then**
14:                 $(\mathcal{V}_{\text{group},j'})_{j'} \leftarrow (\mathcal{V}_{\text{group},j'})_{j'} \cup \{\{v^{i,e}\}\}$
15:             **end if**
16:         **end if**
17:     **end for**
18:     **for** $\mathcal{V}_{\text{group},j}$ in $(\mathcal{V}_{\text{group},j'})_{j'}$ **do**
19:         **if** $\mathcal{V}_{\text{group},j}$ is complete **then**
20:             Transform the segments in $\mathcal{V}_{\text{group},j}$ into a mode $v_{j,e}$
21:             $\mathcal{V} \leftarrow \mathcal{V} \cup \{v_{j,e}\}$
22:             $\mathcal{V}_{\text{seg}} \leftarrow \mathcal{V}_{\text{seg}} \smallsetminus \mathcal{V}_{\text{group},j}$
23:         **end if**
24:     **end for**
25:     **return** $\mathcal{V}, \mathcal{V}_{\text{seg}}$
26: **end function**

$\mathcal{V}_{\text{group}}$ and $\mathcal{V}_{\text{seg}}$. All partition blocks that are not complete will be passed on to the next iteration. These selection steps depend on the prior information about the modes composing the signal and may be based on (a) user input and/or (b) a set of pre-defined rules. Further details and rationale on the options to discard, place segments into partition blocks, and determine the completeness of a block, will be discussed in the following paragraphs. The first loop in Algorithm 6, lines 3 to 17, takes each segment $v^{i,e}$ in $\mathcal{V}_{\text{seg}}$, and either discards it, adds it to a partition block in $\mathcal{V}_{\text{group}}$, or creates a new partition block with it. On line 4, we specify that a segment is to be discarded (i.e., removed from the set of segments $\mathcal{V}_{\text{seg}}$) whenever it corresponds to a mode intersection or noise, where we identify a mode intersection whenever two modes' instantaneous frequencies match at any particular time. This can be seen in Fig. 8.5.1 where the energies for the higher two frequency modes on $t \gtrsim -.25$ meet in frequency at time $t \approx 0.25$, as well as Fig. 8.3.1, where the lower two frequency modes on $t \gtrsim -0.25$ also meet around $t \approx 0.25$. Moreover, segment 0 in Fig. 8.3.3 corresponds to an artifact of this mode intersection. In these two examples, it has been observed selecting ϵ_3 large enough leads to no identified noise artifacts. However, identified segments with these similar characteristics as those in Fig. 8.4.2, i.e., short with rapidly varying frequency, are discarded, especially if there is prior knowledge of noise in the signal.

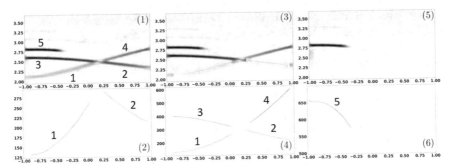

Figure 8.5:   The scale of the vertical axis is $\log_{10}(\omega)$ in the top row of sub-figures (1,3,5) and ω in bottom row of sub-figures (2,4,6). Segments are labeled in (1). (1, 2) Energy $\mathcal{S}(\cdot, \cdot, v)$ and the identified lowest frequency segments (3, 4) First updated energy $\mathcal{S}(\cdot, \cdot, v - v^{1,e} - v^{2,e})$ and its identified lowest frequency segments (5, 6) Second updated energy $\mathcal{S}(\cdot, \cdot, v - v_{1,e} - v_{2,e})$ and its identified lowest frequency segments, where $v_{1,e}$ results from joining mode segments 1 and 4, while $v_{2,e}$ is generated from joining segments 3 and 2.

All segments $v^{i,e}$ that are not discarded are iteratively put into existing partition blocks in lines 7–12 of Algorithm 6, or used to create a new partition block in line 14, which we denote by $\{\{v^{i,e}\}\}$. For example, in Fig. 8.5.2, we place segment 1 into its own partition block on line 14 by default since when $\mathcal{V}_{\text{group}}$ is empty, the loop from lines 7–12 is not executed. Then we do not place segment 2 in the partition block with segment 1, but again place it in its own partition block on line 14 with the observation they belong to different modes (based on the max-squeezed energy $\mathcal{S}$ in Fig. 8.5.1). The end result of this iteration is segments 1 and 2 placed into separate partition blocks. In the next iteration shown in 8.5.4, we construct two partition blocks, one consisting of

$\{\{v^{1,e}, v^{4,e}\}\}$ and the other $\{\{v^{2,e}, v^{3,e}\}\}$. In the following iteration, illustrated in Fig. 8.5.6, we again place segment 5 into its own partition block on line 14 by default. The next iteration is the last since no segments which violate (8.2.7) are observed. Both blocks are then designated as *complete modes*, that is correspond to a mode at all time $[-1, 1]$, and are used to construct $v_{1,e}$ and $v_{2,e}$. This determination can be based on (a) user input and/or (b) a set of pre-defined rules. Observing $\mathcal{S}$ at the third stage in Fig. 8.5.5, we designate it as complete.

The final loop of Algorithm 6 on lines 18–24 begins by checking whether the block is complete. For a block deemed complete, in line 20, their segments are combined to create an estimate of their corresponding mode by interpolating the amplitude and phase to fill the gaps and extrapolation by zero to the boundary. Then, in line 21, this estimated mode is added to $\mathcal{V}$ and, in line 22, its generating segments removed from $\mathcal{V}_{\text{seg}}$. Finally, the segments of the incomplete blocks constitute the output $\mathcal{V}_{\text{seg}}$ of Algorithm 6.

In the implementation corresponding to Fig. 8.5.2, each block consisting of segments 1 and 2, respectively, are both determined to not be complete, and hence are passed to the next iteration as members of $\mathcal{V}_{\text{seg}}$ to the next iteration. In Fig. 8.5.4, the block consisting of segments 1 and 4 and the block consisting of segments 2 and 3 are deemed complete since each block appears to contain different portions of the same mode (with missing portions corresponding to the intersection between the corresponding modes around $t \approx 0.25$), and consequently their segments are therefore designated to be turned into modes $v_{1,e}$ from segments 1 and 4 and $v_{2,e}$ from 2 and 3. Finally, in Fig. 8.5.6, the block consisting of only segment 5 is determined to be complete and in line 20 is extrapolated by zero to produce its corresponding mode. In Example 8.1.2, shown in Fig. 8.3.3, we place segments 1, 2, and 3 in separate blocks (and disregard 0), but only designate the block containing segment 1 as complete. The output of Algorithm 6, and hence Algorithm 4, is the updated list of modes and segments.

8.3   The Segmented Micro-Local KMD Algorithm

The segmented iterated micro-local algorithm identifies full modes in the setting of Problem 5 and is presented in Algorithm 7. Except for the call of the function m_{mode}, Algorithm 4, Algorithm 7 is similar to Algorithm 2. It is initialized by $\mathcal{V} = \varnothing$ and $\mathcal{V}_{\text{seg}} = \varnothing$, and the main iteration between lines 2 and 17 identifies the modes or segments with lowest instantaneous frequency and then provides refined estimates for the amplitude and the phase of each mode $v_i, i \in \{1, \dots, m\}$ of the signal v. We first apply m_{mode} to identify segments to be passed on to the next iteration and mode segments to be combined into modes. This set of recognized modes $\mathcal{V}^{\text{out}}$ will be refined in the loop between lines 8 to 14 by iteratively applying the micro-local KMD steps of Sect. 6.1 on the base frequency of each mode (these steps correspond to the final optimization loop, i.e., lines 21 to 27 in Algorithm 2). The loop is terminated when no additions are made to $\mathcal{V}$ or $\mathcal{V}_{\text{seg}}$.

Algorithm 7 Segmented iterated micro-local KMD

1: $\{\mathcal{V}, \mathcal{V}_{\text{seg}}\} \leftarrow \{\varnothing, \varnothing\}$
2: **while** true **do**
3:     $\{\mathcal{V}^{\text{out}}, \mathcal{V}_{\text{seg}}^{\text{out}}\} \leftarrow m_{\text{mode}}(v, \mathcal{V}, \mathcal{V}_{\text{seg}})$
4:     **if** $\mathcal{V}_{\text{seg}}^{\text{out}} = \varnothing$ and $|\mathcal{V}^{\text{out}}| = |\mathcal{V}|$ **then**
5:         break loop
6:     **end if**
7:     **if** $|\mathcal{V}^{\text{out}}| > |\mathcal{V}|$ **then**
8:         **repeat**
9:             **for** $v_{i,e}$ in $\mathcal{V}^{\text{out}}$ **do**
10:                 $v_{i,\text{res}} \leftarrow v - a_{i,e}\bar{y}(\theta_{i,e}) - \sum_{j\neq i} a_{j,e}y(\theta_{j,e})$
11:                 $a_{i,e}(\tau) \leftarrow a(\tau, \theta_{i,e}, v_{i,\text{res}})/c_1$
12:                 $\theta_{i,e}(\tau) \leftarrow \theta_{i,e}(\tau) + \frac{1}{2}\delta\theta(\tau, \theta_{i,e}, v_{i,\text{res}})$
13:             **end for**
14:         **until** $\sup_{i,\tau} |\delta\theta(\tau, \theta_{i,e}, v_{i,\text{res}})| < \epsilon_1$
15:     **end if**
16:     $\{\mathcal{V}, \mathcal{V}_{\text{seg}}\} \leftarrow \{\mathcal{V}^{\text{out}}, \mathcal{V}_{\text{seg}}^{\text{out}}\}$
17: **end while**
18: Return the modes $v_{i,e}(t) \leftarrow a_{i,e}(t)y(\theta_{i,e}(t))$ for $i = 1, ..., m$

8.4 Numerical Experiments

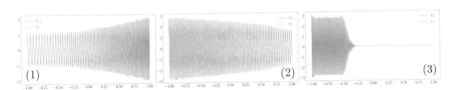

Figure 8.6: (1) $v_{1,e}$ and v_1, (2) $v_{2,e}$ and v_2, (3) $v_{3,e}$ and v_3. See footnote 2.

Figure 8.6 and Table 8.1 show the accuracy of Algorithm 7 in recovering the modes of the signal described in Example 8.1.1, the results for Example 8.1.2 appearing essentially the same, and thereby quantify its robustness to noise, vanishing amplitudes, and crossing frequencies. We again take the mesh spanning $[-1, 1]$ spaced at intervals of size $\frac{1}{5000}$ and aim to recover each mode v_i on the whole interval $[-1, 1]$. We kept $\alpha = 25$ constant in our implementation. The amplitudes and frequencies of the modes composing v are shown in Fig. 8.1. The recovery errors of the modes are found to be consistently on the order of 10^{-2}. Note that in the noise-free setting with identical modes, the recovery error is on the order of 10^{-3} implying the noise is mainly responsible for the errors shown in Table 8.1.

Mode	$\frac{\|v_{i,e}-v_i\|_{L^2}}{\|v_i\|_{L^2}}$	$\frac{\|v_{i,e}-v_i\|_{L^\infty}}{\|v_i\|_{L^\infty}}$	$\frac{\|a_{i,e}-a_i\|_{L^2}}{\|a_i\|_{L^2}}$	$\|\theta_{i,e}-\theta_i\|_{L^2}$
$i=1$	3.17×10^{-2}	6.99×10^{-2}	2.24×10^{-2}	1.99×10^{-2}
$i=2$	2.49×10^{-2}	7.09×10^{-2}	1.64×10^{-2}	1.81×10^{-2}
$i=3$	3.52×10^{-2}	9.52×10^{-2}	3.13×10^{-2}	2.02×10^{-2}

Table 8.1: Signal component recovery errors in Example 8.1.1. Note that the error in phase for mode $i=3$ was calculated over $[-1, -\frac{1}{3}]$ since the phase of a zero signal is undefined.

Appendix

9.1 Universality of the Aggregated Kernel

Let

$$y(t) := \sum_{-N}^{N} c_n e^{int}$$

be the Fourier expansion of a general 2π periodic complex-valued waveform, which we will refer to as the *base waveform*, and use it to define wavelets

$$\chi_{\tau,\omega,\theta}(t) := \omega^{\frac{1-\beta}{2}} y\big(\omega(t-\tau)+\theta\big) e^{-\frac{\omega^2}{\alpha^2}|t-\tau|^2}$$

as in the β-parameterized wavelet versions of (4.5.1) in Theorem 5.3.1, using the waveform y instead of the cosine. The following lemma evaluates the aggregated kernel

$$K_\beta(s,t) := \Re \int_{-\pi}^{\pi} \int_{\mathbb{R}_+} \int_{\mathbb{R}} \chi_{\tau,\omega,\theta}(s)\chi_{\tau,\omega,\theta}^*(t)d\tau d\omega d\theta. \tag{9.1.1}$$

Lemma 9.1.1. *Define the norm*

$$\|y\|^2 := \sum_{n=-N}^{N} e^{-\frac{|n|\alpha^2}{2}}|c_n|^2 \tag{9.1.2}$$

of the base waveform y. We have

$$K_\beta(s,t) = 2\pi|s-t|^{\beta-1} \sum_{n=-N}^{N} a_n(s,t)|c_n|^2,$$

where

$$a_n(s,t) = \frac{\alpha\sqrt{\pi}}{2\sqrt{2}}(\sqrt{2}\alpha)^{1-\beta}\Gamma(\frac{1-\beta}{2})e^{-\frac{|n|\alpha^2}{2}} {}_1F_1\left(\frac{\beta}{2};\frac{1}{2};\frac{|n|\alpha^2}{2}\right).$$

In particular, at $\beta = 0$, we have

$$K_0(s,t) = \alpha^2\pi^2|s-t|^{-1}\|y\|^2.$$

© The Author(s), under exclusive license to Springer Nature Switzerland
AG 2021
H. Owhadi et al., *Kernel Mode Decomposition and the Programming of Kernels*, Surveys and Tutorials in the Applied Mathematical Sciences 8,
https://doi.org/10.1007/978-3-030-82171-5_9

9.1.1 Characterizing the Norm $\sum_{n=-N}^{N} e^{-\frac{|n|\alpha^2}{2}} |c_n|^2$

The norm (9.1.2) of the function $y(t) := \sum_{-N}^{N} c_n e^{int}$ is expressed in terms of its Fourier coefficients c_n. The following lemma evaluates it directly in terms of the function y.

Lemma 9.1.2. *The norm* (9.1.2) *of the function* $y(t) := \sum_{-N}^{N} c_n e^{int}$ *satisfies*

$$\|y\|^2 = \int_{-\pi}^{\pi} \int_{-\pi}^{\pi} G(t,t') y(t) y^*(t') \, dt \, dt',$$

where

$$G(t,t') = 2\pi \frac{\sinh(\frac{\alpha^2}{2})}{\cosh(\frac{\alpha^2}{2}) - \cos(t - t')}, \quad t, t' \in [-\pi, \pi].$$

Remark 9.1.3. *The norm* (9.1.2) *is clearly insensitive to the size of the high frequency (large n) components* $c_n e^{int}$ *of y. On the other hand, the alternative representation of this norm in Lemma 9.1.2 combined with the fact that the kernel G satisfies*

$$\frac{\sinh(\frac{\alpha^2}{2})}{\cosh(\frac{\alpha^2}{2}) + 1} \leqslant G(t,t') \leqslant 2\pi \frac{\sinh(\frac{\alpha^2}{2})}{\cosh(\frac{\alpha^2}{2}) - 1}, \quad t, t' \in [-\pi, \pi],$$

which, for $\alpha \geqslant 10$, *implies*

$$1 - 10^{-21} \leqslant G(t,t') \leqslant 1 + 10^{-21}, \quad t, t' \in [-\pi, \pi],$$

implies that

$$\left| \|y\|^2 - \left| \int_{-\pi}^{\pi} y(t) \, dt \right|^2 \right| \leqslant 10^{-21} \left| \int_{-\pi}^{\pi} |y(t)| \, dt \right|^2,$$

that is, $\|y\|^2$ *is exponentially close to the square of its integral.*

9.2 Proofs

9.2.1 Proof of Lemma 3.1.1

We first establish that $\Psi(v) = \Phi^+ v$, where the Moore-Penrose inverse Φ^+ is defined by $\Phi^+ := \Phi^T (\Phi \Phi^T)^{-1}$, where Φ^T is the Hilbert space adjoint of Φ. To that end, let w^* be the solution of (3.1.4). Since $\Phi : \mathcal{B} \to V$ is surjective, it follows that $\Phi : \mathrm{Ker}^{\perp}(\Phi) \to V$ is a bijection, and therefore

$$\{w : \Phi w = v\} = w_0 + \mathrm{Ker}(\Phi)$$

for a unique $w_0 \in \mathrm{Ker}^{\perp}(\Phi)$. Therefore, setting $w' := w - w_0$, we find that $(w')^* := w^* - w_0$ is a solution of

$$\begin{cases} \text{Minimize } \|w' + w_0\|_{\mathcal{B}} \\ \text{Subject to } w' \in \mathcal{B} \text{ and } \Phi w' = 0, \end{cases} \qquad (9.2.1)$$

so that by the projection theorem we have $(w')^* = P_{\text{Ker}(\Phi)}(-w_0)$, where $P_{\text{Ker}(\Phi)}$ is the orthogonal projection onto $\text{Ker}(\Phi)$. Therefore, $w^* = w_0 + (w')^* = w_0 - P_{\text{Ker}(\Phi)}(w_0) = P_{\text{Ker}^\perp(\Phi)}w_0$, so that we obtain

$$w^* = P_{\text{Ker}^\perp(\Phi)}w_0.$$

Since Φ is surjective and continuous, it follows from the closed range theorem, see e.g., Yosida [125, p. 208], that $\text{Im}(\Phi^T) = \text{Ker}^\perp(\Phi)$ and $\text{Ker}(\Phi^T) = \varnothing$, which implies that $\Phi\Phi^T : V \to V$ is invertible, so that the Moore-Penrose inverse $\Phi^+ : V \to \mathcal{B}$ of Φ is well defined by

$$\Phi^+ := \Phi^T\left(\Phi\Phi^T\right)^{-1}.$$

It follows that $P_{\text{Ker}^\perp(\Phi)} = \Phi^+\Phi$ and $\Phi\Phi^+ = I_V$ so that

$$w^* = P_{\text{Ker}^\perp(\Phi)}w_0 = \Phi^+\Phi w_0 = \Phi^+v,$$

that is, we obtain the second assertion $w^* = \Phi^+v$.

For the first assertion, suppose that $\text{Ker}\,\Phi = \varnothing$. Since it is surjective, it follows that Φ is a bijection. Then, the unique solution to the minmax problem is the only feasible one $w^* = \Phi^{-1}v = \Phi^+v$. When $\text{Ker}\,\Phi \neq \varnothing$, observe that since all u that satisfy $\Phi u = v$ have the representation $u = w_0 + u'$ for fixed $w_0 \in \text{Ker}^\perp(\Phi)$ and some $u' \in \text{Ker}(\Phi)$, it follows that the inner maximum satisfies

$$\max_{u \in \mathcal{B}|\Phi u = v} \frac{\|u - w\|_\mathcal{B}}{\|u\|_\mathcal{B}} = \max_{u' \in \text{Ker}(\Phi)} \frac{\|u' + w_0 - w\|_\mathcal{B}}{\|u' + w_0\|_\mathcal{B}}$$

$$= \max_{u' \in \text{Ker}(\Phi)} \max_{t \in \mathbb{R}} \frac{\|tu' + w_0 - w\|_\mathcal{B}}{\|tu' + w_0\|_\mathcal{B}}$$

$$\geqslant 1.$$

On the other hand, for $w := \Phi^+v$, we have

$$\max_{u \in \mathcal{B}|\Phi u = v} \frac{\|u - w\|_\mathcal{B}}{\|u\|_\mathcal{B}} = \max_{u \in \mathcal{B}|\Phi u = v} \frac{\|u - \Phi^+v\|_\mathcal{B}}{\|u\|_\mathcal{B}}$$

$$= \max_{u \in \mathcal{B}|\Phi u = v} \frac{\|u - \Phi^+\Phi u\|_\mathcal{B}}{\|u\|_\mathcal{B}}$$

$$= \max_{u \in \mathcal{B}|\Phi u = v} \frac{\|u - P_{\text{Ker}^\perp(\Phi)}u\|_\mathcal{B}}{\|u\|_\mathcal{B}}$$

$$\leqslant 1,$$

which implies that $w := \Phi^+v$ is a minmax solution. To see that it is the unique optimal solution, observe that we have just established that

$$\max_{u \in \mathcal{B}|\Phi u = v} \frac{\|u - \Psi(v)\|_\mathcal{B}}{\|u\|_\mathcal{B}} = 1 \tag{9.2.2}$$

for any optimal $\Psi : V \to \mathcal{B}$. It then follows that

$$\max_{u \in \mathcal{B}} \frac{\|u - \Psi(\Phi u)\|_\mathcal{B}}{\|u\|_\mathcal{B}} = 1,$$

which implies that the map $I - \Psi \circ \Phi : \mathcal{B} \to \mathcal{B}$ is a contraction. Moreover, by selecting $u \in \mathrm{Ker}(\Phi)$ tending to 0, it follows from (9.2.2) that $\Psi(0) = 0$. Since, by definition, $\Phi \circ \Psi = I_V$, we have

$$
\begin{aligned}
(I - \Psi \circ \Phi)^2(u) &= (I - \Psi \circ \Phi)(u - \Psi \circ \Phi u) \\
&= u - \Psi \circ \Phi u - \Psi \circ \Phi(u - \Psi \circ \Phi u) \\
&= u - \Psi \circ \Phi u - \Psi(\Phi u - \Phi \circ \Psi \circ \Phi u) \\
&= u - \Psi \circ \Phi u - \Psi(\Phi u - \Phi u) \\
&= u - \Psi \circ \Phi u - \Psi(0) \\
&= u - \Psi \circ \Phi u,
\end{aligned}
$$

so that the map $I - \Psi \circ \Phi$ is a projection. Since $\Phi(u - \Psi \circ \Phi u) = \Phi u - \Phi \circ \Psi \circ \Phi u = 0$, it follows that $\mathrm{Im}(I - \Psi \circ \Phi) \subset \mathrm{Ker}(\Phi)$, but since for $b \in \mathrm{Ker}(\Phi)$, we have $(I - \Psi \circ \Phi)(b) = b - \Psi \circ \Phi b = b$, we obtain the equality $\mathrm{Im}(I - \Psi \circ \Phi) = \mathrm{Ker}(\Phi)$.

To show that a projection of this form is necessarily linear, let us demonstrate that $\mathrm{Im}(\Psi \circ \Phi) = \mathrm{Ker}^{\perp}(\Phi)$. To that end, use the decomposition $\mathcal{B} = \mathrm{Ker}(\Phi) \oplus \mathrm{Ker}^{\perp}(\Phi)$ to write $u = u' + u''$ with $u' \in \mathrm{Ker}(\Phi)$ and $u'' \in \mathrm{Ker}^{\perp}(\Phi)$ and write the contractive condition $\|u - \Psi \circ \Phi u\|^2 \leqslant \|u\|^2$ as

$$
\|u' + u'' - \Psi \circ \Phi(u' + u'')\|^2 \leqslant \|u' + u''\|^2,
$$

which using the linearity of Φ and $u' \in \mathrm{Ker}(\Phi)$ we obtain

$$
\|u' + u'' - \Psi \circ \Phi u''\|^2 \leqslant \|u' + u''\|^2.
$$

Suppose that $\Psi \circ \Phi u'' = v' + v''$ with $v' \in \mathrm{Ker}(\Phi)$ nontrivial. Then, selecting $u' = tv'$, with $t \in \mathbb{R}$, we obtain

$$
\|(t - 1)v' + u'' - v''|^2 \leqslant \|tv' + u''\|^2,
$$

which amounts to

$$
(t - 1)^2 \|v'\|^2 + \|u'' - v''|^2 \leqslant t^2 \|v'\|^2 + \|u''\|^2,
$$

and therefore

$$
(1 - 2t)\|v'\|^2 + \|u'' - v''|^2 \leqslant \|u''\|^2,
$$

which provides a contradiction for t large enough negative. Consequently, $v' = 0$ and $\mathrm{Im}(\Psi \circ \Phi) \subset \mathrm{Ker}^{\perp}(\Phi)$. Since $I = \Psi \circ \Phi + (I - \Psi \circ \Phi)$ with $\mathrm{Im}(\Psi \circ \Phi) \subset \mathrm{Ker}^{\perp}(\Phi)$ and $\mathrm{Im}(I - \Psi \circ \Phi) \subset \mathrm{Ker}(\Phi)$, it follows that $\mathrm{Im}(\Psi \circ \Phi) = \mathrm{Ker}^{\perp}(\Phi)$. Since $\Psi \circ \Phi$ is a projection, it follows that

$$
\Psi \circ \Phi u'' = u'', \quad u'' \in \mathrm{Ker}^{\perp}(\Phi).
$$

Consequently, for two elements $u_1 = u_1' + u_1''$ and $u_2 = u_2' + u_2''$ with $u_i' \in \text{Ker}(\Phi)$ and $u_i'' \in \text{Ker}^\perp(\Phi)$ for $i = 1, 2$ we have

$$
\begin{aligned}
\big(I - \Psi \circ \Phi\big)(u_1 + u_2) &= u_1 + u_2 - \Psi \circ \Phi(u_1 + u_2) \\
&= u_1' + u_2' + u_1'' + u_2'' - \Psi \circ \Phi(u_1'' + u_2'') \\
&= u_1' + u_2' \\
&= u_1' + u_1'' - \Psi \circ \Phi u_1'' + u_2' + u_2'' - \Psi \circ \Phi u_2'' \\
&= \big(I - \Psi \circ \Phi\big)(u_1) + \big(I - \Psi \circ \Phi\big)(u_2),
\end{aligned}
$$

and similarly, for $t \in \mathbb{R}$,

$$
\big(I - \Psi \circ \Phi\big)(t u_1) = t\big(I - \Psi \circ \Phi\big)(u_1),
$$

so we conclude that $I - \Psi \circ \Phi$ is linear.

Since according to Rao [92, Rem. 9, p. 51], a contractive linear projection on a Hilbert space is an orthogonal projection, it follows that the map $I - \Psi \circ \Phi$ is an orthogonal projection, and therefore $\Psi \circ \Phi = P_{\text{Ker}^\perp(\Phi)}$. Since Φ^+ is the Moore-Penrose inverse, it follows that $P_{\text{Ker}^\perp(\Phi)} = \Phi^+ \Phi$ so that $\Psi \circ \Phi = \Phi^+ \Phi$, and therefore the assertion $\Psi = \Phi^+$ follows by right multiplication by Ψ using the identity $\Phi \circ \Psi = I_V$.

9.2.2 Proof of Lemma 3.1.2

Let us write $\Phi : \mathcal{B} \to V$ as

$$
\Phi u = \sum_{i \in \mathcal{I}} e_i u_i, \quad u = (u_i \in V_i)_{i \in \mathcal{I}},
$$

where we now include the subspace injections $e_i : V_i \to V$ in its description. Let $\bar{e}_i : V_i \to \mathcal{B}$ denote the component injection $\bar{e}_i v_i := (0, \ldots, 0, v_i, 0, \ldots, 0)$, and let $\bar{e}_i^T : \mathcal{B} \to V_i$ denote the component projection. Using this notation, the norm (3.1.5) on $\mathcal{B}$ becomes

$$
\|u\|_{\mathcal{B}}^2 := \sum_{i \in \mathcal{I}} \|\bar{e}_i^T u\|_{V_i}^2, \qquad u \in \mathcal{B}, \tag{9.2.3}
$$

with inner product

$$
\big\langle u_1, u_2 \big\rangle_{\mathcal{B}} := \sum_{i \in \mathcal{I}} \big\langle \bar{e}_i^T u_1, \bar{e}_i^T u_2 \big\rangle_{V_i}, \qquad u_1, u_2 \in \mathcal{B}.
$$

Clearly, $\bar{e}_j^T \bar{e}_i = 0, i \neq j$ and $\bar{e}_i^T \bar{e}_i = I_{V_i}$, so that

$$
\begin{aligned}
\big\langle \bar{e}_i^T u, v_i \big\rangle_{V_i} &= \big\langle \bar{e}_i^T u, \bar{e}_i^T \bar{e}_i v_i \big\rangle_{V_i} \\
&= \sum_{j \in \mathcal{I}} \big\langle \bar{e}_j^T u, \bar{e}_j^T \bar{e}_i v_i \big\rangle_{V_i} \\
&= \big\langle u, \bar{e}_i v_i \big\rangle_{\mathcal{B}}
\end{aligned}
$$

implies that $\bar{e}_i^T$ is indeed the adjoint of $\bar{e}_i$. Consequently, we obtain

$$\Phi = \sum_{i \in \mathcal{I}} e_i \bar{e}_i^T,$$

and therefore its Hilbert space adjoint $\Phi^T : V \to \mathcal{B}$ is

$$\Phi^T = \sum_{i \in \mathcal{I}} \bar{e}_i e_i^T,$$

where $e_i^T : V \to V_i$ is the Hilbert space adjoint of e_i. To compute it, use the Riesz isomorphism

$$\iota : V \to V^*$$

and the usual duality relationships to obtain

$$e_i^T = Q_i e_i^* \iota,$$

where $e_i^* : V^* \to V_i^*$ is the dual adjoint projection. Consequently, we obtain

$$\Phi \Phi^T = \sum_{j \in \mathcal{I}} e_j \bar{e}_j^T \sum_{i \in \mathcal{I}} \bar{e}_i e_i^T = \sum_{i,j \in \mathcal{I}} e_j \bar{e}_j^T \bar{e}_i e_i^T = \sum_{i \in \mathcal{I}} e_i e_i^T = \sum_{i \in \mathcal{I}} e_i Q_i e_i^* \iota,$$

and therefore defining

$$S := \sum_{i \in \mathcal{I}} e_i Q_i e_i^*,$$

it follows that

$$\Phi \Phi^T = S \iota.$$

Since $\Phi \Phi^T$ and ι are invertible, S is invertible. The invertibility of S implies both assertions regarding norms and their duality follow in a straightforward way from the definition of the dual norm. For the Hilbert space version, see e.g., [82, Prop. 11.4].

9.2.3 Proof of Theorem 3.1.3

We use the notations and results in the proof of Lemma 3.1.2. The assumption $V = \sum_i V_i$ implies that the information map $\Phi : \mathcal{B} \to V$ defined by

$$\Phi u = \sum_{i \in \mathcal{I}} u_i, \quad u = (u_i \in V_i)_{i \in \mathcal{I}},$$

is surjective. Consequently, Lemma 3.1.1 asserts that the minimizer of (3.1.4) is $w^* = \Psi(v) := \Phi^+ v$, where the Moore-Penrose inverse $\Phi^+ := \Phi^T (\Phi \Phi^T)^{-1}$ of Φ is well defined, with $\Phi^T : V \to \mathcal{B}$ being the Hilbert space adjoint to $\Phi : \mathcal{B} \to V$. The proof of Lemma 3.1.2 obtained $\Phi \Phi^T = S \iota$, where $S := \sum_{i \in \mathcal{I}} e_i Q_i e_i^*$ and $\iota : V \to V^*$ is the Riesz isomorphism, $e_i^T = Q_i e_i^* \iota$, where $e_i^T : V \to V_i$ is the Hilbert space adjoint of e_i and $e_i^* : V^* \to V_i^*$ is its dual space adjoint, and $\Phi^T = \sum_{i \in \mathcal{I}} \bar{e}_i e_i^T$, where $\bar{e}_i : V_i \to \mathcal{B}$ denotes the component injection $\bar{e}_i v_i := (0, \ldots, 0, v_i, 0, \ldots, 0)$.

Therefore, since $(\Phi\Phi^T)^{-1} = \iota^{-1}S^{-1}$, we obtain $\Phi^+ = \sum_{i\in\mathcal{I}} \bar{e}_i Q_i e_i^* \iota \iota^{-1} S^{-1}$, which amounts to

$$\Phi^+ = \sum_{i\in\mathcal{I}} \bar{e}_i Q_i e_i^* S^{-1}, \tag{9.2.4}$$

or in coordinates

$$(\Phi^+ v)_i = Q_i e_i^* S^{-1} v, \quad i \in \mathcal{I},$$

establishing the first assertion. The second follows from the general property $\Phi\Phi^+ = \Phi\Phi^T(\Phi\Phi^T)^{-1} = I$ of the Moore-Penrose inverse. The first isometry assertion follows from

$$
\begin{aligned}
\|\Phi^+ v\|_{\mathcal{B}}^2 &= \sum_{i\in\mathcal{I}} \|(\Phi^+ v)_i\|_{V_i}^2 = \sum_{i\in\mathcal{I}} \|Q_i e_i^* S^{-1} v\|_{V_i}^2 \\
&= \sum_{i\in\mathcal{I}} [Q_i^{-1} Q_i e_i^* S^{-1} v, Q_i e_i^* S^{-1} v] = \sum_{i\in\mathcal{I}} [e_i^* S^{-1} v, Q_i e_i^* S^{-1} v] \\
&= \sum_{i\in\mathcal{I}} [S^{-1} v, e_i Q_i e_i^* S^{-1} v] = [S^{-1} v, \sum_{i\in\mathcal{I}} e_i Q_i e_i^* S^{-1} v] \\
&= [S^{-1} v, S S^{-1} v] = [S^{-1} v, v] = \|v\|_{S^{-1}}^2
\end{aligned}
$$

for $v \in V$.

For the second, write $\Phi = \sum_{i\in\mathcal{I}} e_i \bar{e}_i^T$ and consider its dual space adjoint $\Phi : V^* \to \mathcal{B}^*$ defined by

$$\Phi^* = \sum_{i\in\mathcal{I}} \bar{e}_i^{T,*} e_i^* .$$

A straightforward calculation shows that $\bar{e}_i^{T,*} : V_i^* \to \mathcal{B}^*$ is the component injection into the product $\mathcal{B}^* = \prod_{i\in\mathcal{I}} V_i^*$. Consequently, we obtain

$$\bar{e}_i^T Q \bar{e}_j^{T,*} = \delta_{i,j} Q_j, \quad i, j \in \mathcal{I},$$

so that

$$\Phi Q \Phi^* = \sum_{i\in\mathcal{I}} e_i \bar{e}_i^T Q \sum_{j\in\mathcal{I}} \bar{e}_j^{T,*} e_j^* = \sum_{i,j\in\mathcal{I}} e_i \bar{e}_i^T Q \bar{e}_j^{T,*} e_j^* = \sum_{i\in\mathcal{I}} e_i Q_i e_i^* = S,$$

and since, for $\phi \in V^*$,

$$\|\Phi^* \phi\|_{\mathcal{B}}^2 = \langle \Phi^* \phi, \Phi^* \phi \rangle_{\mathcal{B}*} = [\Phi^* \phi, Q\Phi^* \phi] = [\phi, \Phi Q \Phi^* \phi] = [\phi, S\phi] = \|\phi\|_S^2,$$

it follows that Φ^* is an isometry.

9.2.4 Proof of Theorem 3.1.4

Use the Riesz isomorphism between V and V^* to represent the dual space adjoint $\Phi^* : V^* \to \mathcal{B}^*$ of $\Phi : \mathcal{B} \to V$ as $\Phi^* : V \to \mathcal{B}^*$. It follows from the definition of the Hilbert space adjoint $\Phi^T : V \to \mathcal{B}$ that

$$[\Phi^* v, b] = \langle v, \Phi b \rangle = \langle \Phi^T v, b \rangle_{\mathcal{B}}.$$

Since $Q : \mathcal{B}^* \to \mathcal{B}$ (3.1.10) defines the $\mathcal{B}$ inner product through

$$\langle b_1, b_2 \rangle_{\mathcal{B}} = [Q^{-1} b_1, b_2], \quad b_1, b_2 \in \mathcal{B},$$

it follows that $[\Phi^* v, b] = \langle Q \Phi^* v, b \rangle_{\mathcal{B}}$ and therefore $\langle Q \Phi^* v, b \rangle_{\mathcal{B}} = \langle \Phi^T v, b \rangle_{\mathcal{B}}, v \in V, b \in \mathcal{B}$, so we conclude that

$$\Phi^T = Q \Phi^*.$$

Since Theorem 3.1.3 demonstrated that Ψ is the Moore-Penrose inverse Φ^+, which implies that $\Psi \circ \Phi$ is the orthogonal projection onto $\mathrm{Im}(\Phi^T)$, it follows that $\Psi \circ \Phi u \in \mathrm{Im}(\Phi^T)$. However, the identity $\Phi^T = Q \Phi^*$ implies that $\mathrm{Im}(\Phi^T) = Q \, \mathrm{Im}(\Phi^*)$ so that we obtain the first part

$$\| u - \Psi(\Phi u) \|_{\mathcal{B}} = \inf_{\phi \in V^*} \| u - Q \Phi^*(\phi) \|_{\mathcal{B}}$$

of the assertion. The second half follows from the definition (3.1.5) of $\| \cdot \|_{\mathcal{B}}$.

9.2.5   Proof of Proposition 4.1.1

Restating the assertion using the injections $e_i : V_i \to V$, our objective is to establish that

$$E(i) = \mathrm{Var}\left([\phi, e_i \xi_i] \right) = \mathrm{Var}\left(\langle e_i \xi_i, v \rangle_{S^{-1}} \right).$$

Since $[\phi, e_i \xi_i] = [e_i^* \phi, \xi_i]$, it follows that $[\phi, e_i \xi_i] \sim \mathcal{N}(0, [e_i^* \phi, Q_i e_i^* \phi])$ so that $\mathrm{Var}\left([\phi, e_i \xi_i] \right) = [e_i^* \phi, Q_i e_i^* \phi]$, which using $\phi = S^{-1} v$ becomes

$$\mathrm{Var}\left([\phi, e_i \xi_i] \right) = [S^{-1} v, e_i Q_i e_i^* S^{-1} v].$$

On the other hand, the definitions (4.1.1) of $E(i)$, (3.1.6) of $\| \cdot \|_{V_i}$, and Theorem 3.1.3 imply that

$$E(i) := \| \Psi_i(v) \|_{V_i}^2 = [Q_i^{-1} \Psi_i(v), \Psi_i(v)]$$

$$= [Q_i^{-1} Q_i e_i^* S^{-1} v, Q_i e_i^* S^{-1} v] = [e_i^* S^{-1} v, Q_i e_i^* S^{-1} v] = [S^{-1} v, e_i Q_i e_i^* S^{-1} v],$$

so that we conclude the first part $E(i) = \mathrm{Var}\left([\phi, e_i \xi_i] \right)$ of the assertion. Since $[\phi, e_i \xi_i] = [S^{-1} v, e_i \xi_i] = \langle v, e_i \xi_i \rangle_{S^{-1}}$, we obtain the second.

9.2.6   Proof of Theorem 4.3.3

Fix $1 \leqslant k < r \leqslant q$. To apply Theorem 3.1.3, we select $\mathcal{B} := \mathcal{B}^{(k)}$ and $V := \mathcal{B}^{(r)}$ and endow them with the external direct sum vector space structure of products of vector spaces. Since the information operator $\Phi^{(r,k)} : \mathcal{B}^{(k)} \to \mathcal{B}^{(r)}$ defined in (4.3.5) is diagonal with components $\Phi_j^{(r,k)} : \mathcal{B}_j^{(k)} \to V_j^{(r)}, j \in \mathcal{I}^{(r)}$, and the norm on $\mathcal{B}^{(k)} = \prod_{i \in \mathcal{I}^{(r)}} \mathcal{B}_i^{(k)}$ is the product norm $\| u \|_{\prod_{i \in \mathcal{I}^{(r)}} \mathcal{B}_i^{(k)}}^2 =$

$\sum_{i \in \mathcal{I}^{(r)}} \|u_i\|^2_{\mathcal{B}_i^{(k)}}$, $u = (u_i)_{i \in \mathcal{I}^{(r)}}$, it follows from the variational characterization of Lemma 3.1.1, the diagonal nature of the information map $\Phi^{(r,k)}$, and the product metric structure on $\mathcal{B}^{(k)}$ that the optimal recovery solution $\Psi^{(k,r)}$ is the diagonal operator with components the optimal solution operators $\Psi_j^{(k,r)}$, evaluated below, corresponding to the component information maps $\Phi_j^{(r,k)}$: $\mathcal{B}_j^{(k)} \to V_j^{(r)}, j \in \mathcal{I}^{(r)}$. Since each component (4.3.4) of the observation operator is

$$\Phi_j^{(r,k)}(u) := \sum_{i \in j^{(k)}} u_i, \qquad u \in \mathcal{B}_j^{(k)},$$

it follows that the appropriate subspaces of $V_j^{(r)}$ are

$$V_i^{(k)} \subset V_j^{(r)}, \qquad i \in j^{(k)}.$$

Moreover, Condition 4.3.2 and the semigroup nature of the hierarchy of subspace embeddings imply that

$$e_{j,i}^{(k+2,k)} = \sum_{l \in j^{(k+1)}} e_{j,l}^{(k+2,k+1)} e_{l,i}^{(k+1,k)}, \qquad i \in j^{(k)},$$

where the sum, despite its appearance, is over one term, and by induction, we can establish that assumption (4.3.15) implies that

$$Q_j^{(r)} = \sum_{i \in j^{(k)}} e_{j,i}^{(r,k)} Q_i^{(k)} e_{i,j}^{(k,r)}, \qquad j \in \mathcal{I}^{(r)}. \tag{9.2.5}$$

Utilizing the adjoint $e_{i,j}^{(k,r)}$: $V_j^{(r),*} \to V_i^{(k),*}$ (4.3.9) to the subspace embedding $e_{j,i}^{(r,k)}$: $V_i^{(k)} \to V_j^{(r)}$, it now follows from Theorem 3.1.3 and (9.2.5) that these component optimal solution maps $\Psi_j^{(k,r)}$: $V_j^{(r)} \to \mathcal{B}_j^{(k)}$ are those assumed in the theorem in (4.3.10) and (4.3.11) as

$$\Psi_j^{(k,r)}(v_j) := \left(Q_i^{(k)} e_{i,j}^{(k,r)} Q_j^{(r),-1} v_j \right)_{i \in j^{(k)}}, \qquad v_j \in V_j^{(r)}. \tag{9.2.6}$$

The first three assertions for each component j then follow from Theorem 3.1.3, thus establishing the first three assertions in full.

For the semigroup assertions, Condition 4.3.2 implies that, for $k < r < s$ and $l \in \mathcal{I}^{(s)}$, there is a one to one relationship between $\{j \in l^{(r)}, i \in j^{(k)}\}$ and $\{i \in l^{(k)}\}$. Consequently, the definition (4.3.5) of $\Phi^{(r,k)}$ implies

$$\Phi^{(s,r)} \circ \Phi^{(r,k)}(u) = \left(\sum_{j \in l^{(r)}} \left(\sum_{i \in j^{(k)}} u_i \right) \right)_{l \in \mathcal{I}^{(s)}} = \left(\sum_{i \in l^{(k)}} u_i \right)_{l \in \mathcal{I}^{(s)}} = \Phi^{(s,k)}(u),$$

establishing the fourth assertion $\Phi^{(s,k)} = \Phi^{(s,r)} \circ \Phi^{(r,k)}$.

For the fifth, the definition (4.3.12) of $\Psi^{(k,r)}$ implies that

$$
\begin{aligned}
\Psi^{(k,r)} \circ \Psi^{(r,s)}(v) &= \left(Q_i^{(k)} e_{i,j}^{(k,r)} Q_j^{(r),-1} \Psi_j^{(r,s)}(v) \right)_{i \in j^{(k)}} \\
&= \left(Q_i^{(k)} e_{i,j}^{(k,r)} Q_j^{(r),-1} Q_j^{(r)} e_{j,l}^{(r,s)} Q_l^{(s),-1} v_l \right)_{i \in j^{(k)}} \\
&= \left(Q_i^{(k)} e_{i,j}^{(k,r)} e_{j,l}^{(r,s)} Q_l^{(s),-1} v_l \right)_{i \in j^{(k)}} \\
&= \left(Q_i^{(k)} e_{i,l}^{(k,s)} Q_l^{(s),-1} v_l \right)_{i \in l^{(k)}} \\
&= \Psi^{(k,s)}(v),
\end{aligned}
$$

establishing $\Psi^{(k,s)} = \Psi^{(k,r)} \circ \Psi^{(r,s)}$.

The last assertion follows directly from the second and the fifth.

9.2.7   Proof of Theorem 4.3.5

Since $\xi^{(k)} : \mathcal{B}^{(k),*} \to \mathbf{H}$ is an isometry to a Gaussian space of real variables, we can abuse notation and write $\xi^{(k)}(b^*) = [b^*, \xi^{(k)}]$, which emphasizes the interpretation of $\xi^{(k)}$ as a weak $\mathcal{B}^{(k)}$-valued random variable. Since, by Theorem 4.3.3,

$$
\Phi^{(k,1),*} : (\mathcal{B}^{(k),*}, \|\cdot\|_{\mathcal{B}^{(k),*}}) \to (\mathcal{B}^{(1),*}, \|\cdot\|_{\mathcal{B}^{(1),*}}) \text{ is an isometry} \qquad (9.2.7)
$$

and $\xi^{(1)} : \mathcal{B}^{(1),*} \to \mathbf{H}$ is an isometry, it follows that

$$
\Phi^{(k,1)} \xi^{(1)} := \xi^{(1)} \circ \Phi^{(k,1),*} : \mathcal{B}^{(k),*} \to \mathbf{H}
$$

is an isometry and therefore a Gaussian field on $\mathcal{B}^{(k)}$. Since Gaussian fields transform like Gaussian measures with respect to continuous linear transformations, we obtain that $\xi^{(1)} \sim \mathcal{N}(0, Q^1)$ implies that

$$
\Phi^{(k,1)} \xi^{(1)} \sim \mathcal{N}(0, \Phi^{(k,1)} Q^1 \Phi^{(k,1),*}),
$$

but the isometric nature (9.2.7) of $\Phi^{(k,1),*}$ implies that

$$
\Phi^{(k,1)} Q^{(1)} \Phi^{(k,1),*} = Q^{(k)},
$$

so we conclude that

$$
\Phi^{(k,1)} \xi^{(1)} \sim \mathcal{N}(0, Q^k)
$$

thus establishing the assertion that $\xi^{(k)}$ is distributed as $\Phi^{(k,1)} \xi^{(1)}$.

The conditional expectation $\mathbb{E}\big[\xi^{(k)} \mid \Phi^{(r,k)}(\xi^{(k)})\big]$ is uniquely characterized by its field of conditional expectations $\mathbb{E}\big[[b^*, \xi^{(k)}] \mid \Phi^{(r,k)}(\xi^{(k)})\big], b^* \in \mathcal{B}^{(k),*}$, which, because of the linearity of conditional expectation of Gaussian random variables, appears as

$$
\mathbb{E}\big[[b^*, \xi^{(k)}] \mid \Phi^{(r,k)}(\xi^{(k)})\big] = [A_{b*}, \Phi^{(r,k)}(\xi^{(k)})]
$$

for some $A_{b*} \in V^*$. Furthermore, the Gaussian conditioning also implies that the dependence of A_{b*} on b^* is linear, so we write $A_{b*} = Ab^*$ for some $A : \mathcal{B}^* \to V^*$, thereby obtaining

$$\mathbb{E}\big[[b^*, \xi^{(k)}] \mid \Phi^{(r,k)}(\xi^{(k)})\big] = [Ab^*, \Phi^{(r,k)}(\xi^{(k)})], \quad b^* \in \mathcal{B}^{(k),*}. \tag{9.2.8}$$

Using the well-known fact, see e.g., Dudley [25, Thm. 10.2.9], that the conditional expectation of a square-integrable random variable on a probability space (Ω, Σ', P) with respect to a sub-σ-algebra $\Sigma' \subset \Sigma$ is the orthogonal projection onto the closed subspace $L^2(\Omega, \Sigma', P) \subset L^2(\Omega, \Sigma, P)$, it follows that the conditional expectation satisfies

$$\mathbb{E}\big[([b^*, \xi^{(k)}] - [Ab^*, \Phi^{(r,k)}(\xi^{(k)})])[v^*, \Phi^{(r,k)}(\xi^{(k)})]\big] = 0, \quad b^* \in \mathcal{B}^{(k),*}, v^* \in V^{(k),*}.$$

Rewriting this as

$$\mathbb{E}\big[([b^*, \xi^{(k)}] - [\Phi^{(r,k),*}Ab^*, \xi^{(k)}])[\Phi^{(r,k),*}v^*, \xi^{(k)}]\big] = 0, \quad b^* \in \mathcal{B}^{(k),*}, v^* \in V^{(k),*},$$

we obtain

$$\begin{aligned}
[b^*, Q^{(k)}\Phi^{(r,k),*}v^*] &= [\Phi^{(r,k),*}Ab^*, Q^{(k)}\Phi^{(r,k),*}v^*] \\
&= [b^*, A^*\Phi^{(r,k)}Q^{(k)}\Phi^{(r,k),*}v^*]
\end{aligned}$$

for all $b^* \in \mathcal{B}^{(k),*}$ and $v^* \in V^{(k),*}$ and so conclude that

$$A^*\Phi^{(r,k)}Q^{(k)}\Phi^{(r,k),*}v^* = Q^{(k)}\Phi^{(r,k),*}v^*, \quad b^* \in \mathcal{B}^{(k),*}, v^* \in V^{(k),*},$$

which implies that

$$A^*\Phi^{(r,k)}b = b, \quad b \in \mathrm{Im}(Q^{(k)}\Phi^{(r,k),*}). \tag{9.2.9}$$

Since

$$\begin{aligned}
\langle \Phi^{(r,k),T}b^{(r)}, b^{(k)}\rangle_{\mathcal{B}^{(k)}} &= \langle b^{(r)}, \Phi^{(r,k)}b^{(k)}\rangle_{\mathcal{B}^{(k)}} \\
&= [Q^{(k),-1}b^{(r)}, \Phi^{(r,k)}b^{(k)}] \\
&= [\Phi^{(r,k),*}Q^{(k),-1}b^{(r)}, b^{(k)}] \\
&= [Q^{(r),-1}Q^{(r)}\Phi^{(r,k),*}Q^{(k),-1}b^{(r)}, b^{(k)}] \\
&= \langle Q^{(r)}\Phi^{(r,k),*}Q^{(k),-1}b^{(r)}, b^{(k)}\rangle_{\mathcal{B}^{(r)}},
\end{aligned}$$

we conclude that

$$\Phi^{(r,k),T} = Q^{(r)}\Phi^{(r,k),*}Q^{(k),-1},$$

and therefore

$$\mathrm{Im}(Q^{(r)}\Phi^{(r,k),*}) = \mathrm{Im}(\Phi^{(r,k),T}).$$

Consequently, (9.2.9) now reads

$$A^*\Phi^{(r,k)}b = b, \quad b \in \mathrm{Im}(\Phi^{(r,k),T}). \tag{9.2.10}$$

Since clearly

$$A^* \Phi^{(r,k)} b = 0, \quad b \in \mathrm{Ker}(\Phi^{(r,k)}),$$

it follows that

$$A^* \Phi^{(r,k)} = P_{\mathrm{Im}(\Phi^{(r,k),T})}.$$

Since $P_{\mathrm{Im}(\Phi^{(r,k),T})} = (\Phi^{(r,k)})^+ \Phi^{(r,k)}$, the identity $\Phi^{(r,k)} (\Phi^{(r,k)})^+ = I$ establishes that

$$A^* = (\Phi^{(r,k)})^+.$$

Since (9.2.8) implies that

$$\mathbb{E}\big[[b^*, \xi^{(k)}] \mid \Phi^{(r,k)}(\xi^{(k)}) \big] = [b^*, A^* \Phi^{(r,k)}(\xi^{(k)})], \quad b^* \in \mathcal{B}^{(k),*},$$

which in turn implies that

$$\mathbb{E}\big[\xi^{(k)} \mid \Phi^{(r,k)}(\xi^{(k)}) \big] = A^* \Phi^{(r,k)}(\xi^{(k)}),$$

we obtain

$$\mathbb{E}\big[\xi^{(k)} \mid \Phi^{(r,k)}(\xi^{(k)}) \big] = (\Phi^{(r,k)})^+ \Phi^{(r,k)}(\xi^{(k)}).$$

Since Theorem 3.1.3 established that the optimal solution operator $\Psi^{(k,r)}$ corresponding to the information map $\Phi^{(r,k)}$ was the Moore-Penrose inverse $\Psi^{(k,r)} = (\Phi^{(r,k)})^+$, we obtain

$$\mathbb{E}\big[\xi^{(k)} \mid \Phi^{(r,k)}(\xi^{(k)}) \big] = \Psi^{(k,r)} \circ \Phi^{(r,k)}(\xi^{(k)}), \qquad (9.2.11)$$

so that

$$\mathbb{E}\big[\xi^{(k)} \mid \Phi^{(r,k)}(\xi^{(k)}) = v \big] = \Psi^{(k,r)}(v),$$

thus establishing the final assertion. To establish the martingale property, let us define $\hat{\xi}^{(1)} := \xi^{(1)}$ and

$$\hat{\xi}^{(k)} := \mathbb{E}\big[\xi^{(1)} \mid \Phi^{(k,1)}(\xi^{(1)}) \big], \quad k = 2, \ldots$$

as a sequence of Gaussian fields all on the same space $\mathcal{B}^{(1)}$. Equation (9.2.11) implies that

$$\hat{\xi}^{(k)} = \Psi^{(1,k)} \circ \Phi^{(k,1)}(\xi^{(1)}), \qquad (9.2.12)$$

so that the identities $\Phi^{(r,1)} = \Phi^{(r,k)} \circ \Phi^{(k,1)}$ and $\Phi^{(k,1)} \circ \Psi^{(1,k)} = I_{\mathcal{B}^{(1)}}$ from Theorem 4.3.3 imply that

$$
\begin{aligned}
\mathbb{E}[\hat{\xi}^{(k)} | \Phi^{(r,1)}(\hat{\xi}^{(k)})] &= \mathbb{E}[\Psi^{(1,k)} \circ \Phi^{(k,1)}(\hat{\xi}^{(1)}) | \Phi^{(r,1)} \circ \Psi^{(1,k)} \circ \Phi^{(k,1)}(\hat{\xi}^{(1)})] \\
&= \mathbb{E}[\Psi^{(1,k)} \circ \Phi^{(k,1)}(\hat{\xi}^{(1)}) | \Phi^{(r,k)} \circ \Phi^{(k,1)} \circ \Psi^{(1,k)} \circ \Phi^{(k,1)}(\hat{\xi}^{(1)})] \\
&= \mathbb{E}[\Psi^{(1,k)} \circ \Phi^{(k,1)}(\hat{\xi}^{(1)}) | \Phi^{(r,k)} \circ \Phi^{(k,1)}(\hat{\xi}^{(1)})] \\
&= \mathbb{E}[\Psi^{(1,k)} \circ \Phi^{(k,1)}(\hat{\xi}^{(1)}) | \Phi^{(r,1)}(\hat{\xi}^{(1)})] \\
&= \Psi^{(1,k)} \circ \Phi^{(k,1)} \mathbb{E}[\hat{\xi}^{(1)} | \Phi^{(r,1)}(\hat{\xi}^{(1)})] \\
&= \Psi^{(1,k)} \circ \Phi^{(k,1)} \hat{\xi}^{(r)} \\
&= \Psi^{(1,k)} \circ \Phi^{(k,1)} \Psi^{(1,r)} \circ \Phi^{(r,1)} \hat{\xi}^{(1)} \\
&= \Psi^{(1,k)} \circ \Phi^{(k,1)} \Psi^{(1,k)} \Psi^{(k,r)} \circ \Phi^{(r,1)} \hat{\xi}^{(1)} \\
&= \Psi^{(1,k)} \circ \Psi^{(k,r)} \circ \Phi^{(r,1)} \hat{\xi}^{(1)} \\
&= \Psi^{(1,r)} \circ \Phi^{(r,1)} \hat{\xi}^{(1)} \\
&= \hat{\xi}^{(r)},
\end{aligned}
$$

that is, $\hat{\xi}^{(k)}$ is a reverse martingale.

9.2.8   Proof of Theorem 5.3.1

Let us simplify for the moment and define a scaled wavelet

$$
\bar{\chi}_{\tau,\omega,\theta}(t) := \omega^{\frac{1-\beta}{2}} \cos(\omega(t-\tau)+\theta) e^{-\frac{\omega^2(t-\tau)^2}{\alpha^2}}, \qquad t \in \mathbb{R}, \tag{9.2.13}
$$

so that at $\beta = 0$, we have

$$
\chi_{\tau,\omega,\theta} = \left(\frac{2}{\pi^3 \alpha^2}\right)^{\frac{1}{4}} \bar{\chi}_{\tau,\omega,\theta}. \tag{9.2.14}
$$

Since

$$
\begin{aligned}
K(s,t) &:= \int_{-\pi}^{\pi} \int_{\mathbb{R}_+} \int_{\mathbb{R}} \bar{\chi}_{\tau,\omega,\theta}(s) \bar{\chi}_{\tau,\omega,\theta}(t) d\tau\, d\omega\, d\theta \\
&= \int_{-\pi}^{\pi} \int_{\mathbb{R}_+} \int_{\mathbb{R}} \cos(\omega(s-\tau)+\theta) e^{-\frac{\omega^2(s-\tau)^2}{\alpha^2}} \cos(\omega(t-\tau)+\theta) e^{-\frac{\omega^2(t-\tau)^2}{\alpha^2}} d\tau\, \omega^{1-\beta} d\omega\, d\theta \\
&= \int_{-\pi}^{\pi} \int_{\mathbb{R}_+} \int_{\mathbb{R}} \cos(\omega(s-\tau)+\theta) \cos(\omega(t-\tau)+\theta) e^{-\frac{\omega^2(s-\tau)^2}{\alpha^2} - \frac{\omega^2(t-\tau)^2}{\alpha^2}} d\tau\, \omega^{1-\beta} d\omega\, d\theta,
\end{aligned}
$$

the trigonometric identity

$$
\begin{aligned}
&\cos(\omega(s-\tau)+\theta) \cos(\omega(t-\tau)+\theta) \\
&= \Big(\cos(\omega(s-\tau)) \cos\theta - \sin(\omega(s-\tau)) \sin\theta\Big) \Big(\cos(\omega(t-\tau)) \cos\theta - \sin(\omega(t-\tau)) \sin\theta\Big)
\end{aligned}
$$

and the integral identities $\int_{-\pi}^{\pi} \cos^2\theta\, d\theta = \int_{-\pi}^{\pi} \sin^2\theta\, d\theta = \pi$ and $\int_{-\pi}^{\pi} \cos\theta\sin\theta\, d\theta = 0$ imply that

$$K(s,t) = \pi \int_{\mathbb{R}_+} \int_{\mathbb{R}} \Big(\cos\big(\omega(s-\tau)\big)\cos\big(\omega(t-\tau)\big) + \sin\big(\omega(s-\tau)\big)\sin\big(\omega(t-\tau)\big)\Big)$$
$$\times e^{-\frac{\omega^2(s-\tau)^2}{\alpha^2}} e^{-\frac{\omega^2(t-\tau)^2}{\alpha^2}}\, d\tau\, \omega^{1-\beta}\, d\omega$$

so that the cosine subtraction formula implies

$$K(s,t) = \pi \int_{\mathbb{R}_+} \int_{\mathbb{R}} \cos\big(\omega(s-t)\big) e^{-\frac{\omega^2(s-\tau)^2}{\alpha^2}} e^{-\frac{\omega^2(t-\tau)^2}{\alpha^2}}\, d\tau\, \omega^{1-\beta}\, d\omega,$$

which amounts to

$$K(s,t) = \pi\Re \int_{\mathbb{R}_+} \int_{\mathbb{R}} e^{i\omega(s-t)} e^{-\frac{\omega^2(s-\tau)^2}{\alpha^2}} e^{-\frac{\omega^2(t-\tau)^2}{\alpha^2}}\, d\tau\, \omega^{1-\beta}\, d\omega. \qquad (9.2.15)$$

Using the identity

$$e^{-\frac{\omega^2|s-\tau|^2}{\alpha^2}} e^{-\frac{\omega^2|t-\tau|^2}{\alpha^2}} = e^{-\frac{\omega^2}{\alpha^2}\left(2\tau^2 - 2(s+t)\tau\right)} e^{-\frac{\omega^2}{\alpha^2}\left(s^2+t^2\right)}$$

and the integral identity

$$\int e^{-a\tau^2 - 2b\tau}\, d\tau = \sqrt{\frac{\pi}{a}}\, e^{\frac{b^2}{a}}, \qquad a > 0,\ b \in \mathbb{C}, \qquad (9.2.16)$$

with the choice $a := \frac{2\omega^2}{\alpha^2}$ and $b := -\frac{\omega^2}{\alpha^2}(s+t)$, so that $b^2/a = \frac{\omega^2}{2\alpha^2}(s+t)^2$, we can evaluate the integral

$$\int e^{-\frac{\omega^2}{\alpha^2}\left(2\tau^2 - 2(s+t)\tau\right)}\, d\tau = \frac{\alpha}{\omega}\sqrt{\frac{\pi}{2}}\, e^{\frac{\omega^2}{2\alpha^2}(s+t)^2}.$$

Consequently,

$$\begin{aligned}
K(s,t) &= \pi\Re \int e^{i\omega(s-t)} e^{-\frac{\omega^2|s-\tau|^2}{\alpha^2}} e^{-\frac{\omega^2|t-\tau|^2}{\alpha^2}}\, d\tau\, \omega^{1-\beta}\, d\omega \\[4pt]
&= \pi\Re \int e^{i\omega(s-t)} e^{-\frac{\omega^2}{\alpha^2}\left(s^2+t^2\right)} e^{-\frac{\omega^2}{\alpha^2}\left(2\tau^2 - 2(s+t)\tau\right)}\, d\tau\, \omega^{1-\beta}\, d\omega \\[4pt]
&= \pi\Re \int e^{i\omega(s-t)} e^{-\frac{\omega^2}{\alpha^2}\left(s^2+t^2\right)} \left(\int e^{-\frac{\omega^2}{\alpha^2}\left(2\tau^2 - 2(s+t)\tau\right)}\, d\tau \right) \omega^{1-\beta}\, d\omega \\[4pt]
&= \alpha\sqrt{\frac{\pi^3}{2}}\,\Re \int e^{i\omega(s-t)} e^{-\frac{\omega^2}{\alpha^2}\left(s^2+t^2\right)} e^{\frac{\omega^2}{2\alpha^2}(s+t)^2}\, \omega^{-\beta}\, d\omega \\[4pt]
&= \alpha\sqrt{\frac{\pi^3}{2}}\,\Re \int e^{i\omega(s-t)} e^{-\frac{\omega^2}{2\alpha^2}(s-t)^2}\, \omega^{-\beta}\, d\omega \\[4pt]
&= \alpha\sqrt{\frac{\pi^3}{2}} \int \cos\big(\omega(s-t)\big) e^{-\frac{\omega^2}{2\alpha^2}(s-t)^2}\, \omega^{-\beta}\, d\omega,
\end{aligned}$$

that is,

$$K(s,t) = \alpha\sqrt{\frac{\pi^3}{2}} \int \cos(\omega(s-t))e^{-\frac{\omega^2}{2\alpha^2}(s-t)^2}\omega^{-\beta}d\omega. \qquad (9.2.17)$$

Utilizing the integral identity

$$\int_0^\infty x^{\mu-1}e^{-p^2x^2}\cos(ax)dx = \frac{1}{2}p^{-\mu}\Gamma(\frac{\mu}{2})e^{-\frac{a^2}{4p^2}}{}_1F_1\left(-\frac{\mu}{2}+\frac{1}{2},\frac{1}{2};\frac{a^2}{4p^2}\right), \quad a>0, \mu>0,$$
$$(9.2.18)$$

from Gradshteyn and Ryzhik [41, 3.952:8], with $\frac{a^2}{4p^2} = \frac{\alpha^2}{2}$, $p^2 = \frac{|s-t|^2}{2\alpha^2}$, $a := |s-t|$, and $\mu := 1-\beta$, we obtain

$$K(s,t) = \alpha\sqrt{\frac{\pi^3}{2}}\frac{1}{2}(\sqrt{2}\alpha)^{1-\beta}|s-t|^{\beta-1}\Gamma(\frac{1-\beta}{2})e^{-\frac{\alpha^2}{2}}{}_1F_1\left(\frac{\beta}{2},\frac{1}{2};\frac{\alpha^2}{2}\right).$$

Consequently, reintroducing the scaling, (9.2.14) obtains $K_u(s,t) = \left(\frac{2}{\pi^3\alpha^2}\right)^{\frac{1}{2}}$ $K(s,t)$ when $\beta = 0$. To indicate the dependence on β, we define

$$K_\beta(s,t) = \frac{1}{2}(\sqrt{2}\alpha)^{1-\beta}|s-t|^{\beta-1}\Gamma(\frac{1-\beta}{2})e^{-\frac{\alpha^2}{2}}{}_1F_1\left(\frac{\beta}{2},\frac{1}{2};\frac{\alpha^2}{2}\right), \qquad (9.2.19)$$

so that $K_u = K_0$. For fixed α, at the limit $\beta = 0$, we have, recalling that $\Gamma(\frac{1}{2}) = \sqrt{\pi}$,

$$K_0(s,t) = \frac{\sqrt{2\pi}}{2}\alpha|s-t|^{-1}e^{-\frac{\alpha^2}{2}}{}_1F_1\left(0,\frac{1}{2};\frac{\alpha^2}{2}\right),$$

and since ${}_1F_1\left(0,\frac{1}{2};\frac{\alpha^2}{2}\right) = 1$, we obtain

$$K_0(s,t) = \frac{\sqrt{2\pi}}{2}\alpha|s-t|^{-1}e^{-\frac{\alpha^2}{2}}.$$

The scaling constant $H(\beta)$ defined in the theorem satisfies

$$H(\beta) := \frac{1}{2}(\sqrt{2}\alpha)^{1-\beta}\Gamma(1-\frac{\beta}{2})e^{-\frac{\alpha^2}{2}}{}_1F_1\left(\frac{\beta}{2},\frac{1}{2};\frac{\alpha^2}{2}\right)\bar{H}(\beta)$$

with

$$\bar{H}(\beta) := 2^\beta\pi^{\frac{1}{2}}\frac{\Gamma(\frac{\beta}{2})}{\Gamma(1-\frac{\beta}{2})}, \qquad (9.2.20)$$

so that, by (9.2.19), we have

$$\frac{1}{H(\beta)}K_\beta(s,t) = \frac{|s-t|^{\beta-1}}{\bar{H}(\beta)}.$$

Therefore, if we let $\mathcal{K}_\beta$ denote the integral operator

$$(\mathcal{K}_\beta f)(s) := \frac{1}{H(\beta)}\int_{\mathbb{R}} K_\beta(s,t)f(t)dt$$

associated with the kernel K_β scaled by $H(\beta)$, it follows that

$$(\mathcal{K}_\beta f)(s): \frac{1}{\bar{H}(\beta)} \int_{\mathbb{R}} |s-t|^{\beta-1} f(t) dt,$$

namely that it is a scaled version of the integral operator $f \mapsto \int_{\mathbb{R}} |s-t|^{\beta-1} f(t) dt$ corresponding to the Riesz potential $|s-t|^{\beta-1}$. Consequently, according to Helgason [48, Lem. 5.4 & Prop. 5.5], this scaling of the Riesz potential by $\bar{H}(\beta)$ implies the assertions of the theorem.

9.2.9　Proof of Lemma 9.1.1

The outer most integral in the definition (9.1.1) of K_β is

$$
\begin{aligned}
\int_{-\pi}^{\pi} y(\omega(s-\tau)+\theta) y^*(\omega(t-\tau)+\theta) d\theta &= \int_{-\pi}^{\pi} \sum_{-N}^{N} c_n e^{in(\omega(s-\tau)+\theta)} \sum_{-N}^{N} c_m^* e^{-im(\omega(t-\tau)+\theta)} d\theta \\
&= \sum_{n=-N}^{N} \sum_{m=-N}^{N} e^{in\omega(s-\tau)} e^{-im\omega(t-\tau)} c_n c_m^* \int_{-\pi}^{\pi} e^{i(n-m)\theta} d\theta \\
&= 2\pi \sum_{n=-N}^{N} e^{in\omega(s-\tau)} e^{-in\omega(t-\tau)} |c_n|^2 \\
&= 2\pi \sum_{n=-N}^{N} e^{in\omega(s-t)} |c_n|^2,
\end{aligned}
$$

so that

$$K_\beta(s,t) = 2\pi \sum_{n=-N}^{N} K_n(s,t) |c_n|^2,$$

where

$$
\begin{aligned}
K_n(s,t) &= \Re \int e^{in\omega(s-t)} e^{-\frac{\omega^2 |s-\tau|^2}{\alpha^2}} e^{-\frac{\omega^2 |t-\tau|^2}{\alpha^2}} d\tau \omega^{1-\beta} d\omega \\
&= \Re \int e^{in\omega(s-t)} e^{-\frac{\omega^2}{\alpha^2}(s^2+t^2)} e^{-\frac{\omega^2}{\alpha^2}(2\tau^2-2(s+t)\tau)} d\tau \omega^{1-\beta} d\omega \\
&= \Re \int e^{in\omega(s-t)} e^{-\frac{\omega^2}{\alpha^2}(s^2+t^2)} \left(\int e^{-\frac{\omega^2}{\alpha^2}(2\tau^2-2(s+t)\tau)} d\tau\right) \omega^{1-\beta} d\omega \\
&= \alpha\sqrt{\frac{\pi}{2}} \Re \int e^{in\omega(s-t)} e^{-\frac{\omega^2}{\alpha^2}(s^2+t^2)} e^{\frac{\omega^2}{2\alpha^2}(s+t)^2} \omega^{-\beta} d\omega \\
&= \alpha\sqrt{\frac{\pi}{2}} \Re \int e^{in\omega(s-t)} e^{-\frac{\omega^2}{2\alpha^2}(s-t)^2} \omega^{-\beta} d\omega \\
&= \alpha\sqrt{\frac{\pi}{2}} \int \cos(n\omega(s-t)) e^{-\frac{\omega^2}{2\alpha^2}(s-t)^2} \omega^{-\beta} d\omega.
\end{aligned}
$$

Consequently, using the integral identity (9.2.18) with $a = |n| |s-t|$, $\mu = 1-\beta$, $p^2 = \frac{|s-t|^2}{2\alpha^2}$, and therefore $\frac{a^2}{4p^2} = \frac{|n|\alpha^2}{2}$ and $p = \frac{|s-t|}{\sqrt{2}\alpha}$, we conclude that

$$K_n(s,t) = \frac{\alpha\sqrt{\pi}}{2\sqrt{2}} (\sqrt{2}\alpha)^{1-\beta} |s-t|^{\beta-1} \Gamma\left(\frac{1-\beta}{2}\right) e^{-\frac{|n|\alpha^2}{2}} {}_1F_1\left(\frac{\beta}{2}; \frac{1}{2}; \frac{|n|\alpha^2}{2}\right),$$

which does not appear to have a nice dependency on n, except for $\beta = 0$, where $_1F_1\left(0; \frac{1}{2}; \frac{|n|\alpha^2}{2}\right) = 1$ and $\Gamma(\frac{1}{2}) = \sqrt{\pi}$, so that

$$K_n(s,t) = \frac{1}{2}\alpha^2 \pi e^{-\frac{|n|\alpha^2}{2}}|s - t|^{-1},$$

and therefore

$$K_0(s,t) = \alpha^2 \pi^2 \|y\|^2 |s - t|^{-1},$$

when written in terms of the norm $\|y\|^2 := \sum_{n=-N}^{N} e^{-\frac{|n|\alpha^2}{2}}|c_n|^2$.

9.2.10 Proof of Lemma 9.1.2

For $\gamma > 0$, let us evaluate the function

$$\phi(s) := \sum_{n=-\infty}^{\infty} e^{-|n|\gamma}e^{ins} \tag{9.2.21}$$

with Fourier coefficients $\hat{\phi}(n) = e^{-|n|\gamma}$. Since

$$\begin{aligned}
\phi(s) &= \sum_{n=-\infty}^{\infty} e^{-|n|\gamma}e^{ins} \\
&= \sum_{n=1}^{\infty} e^{-n\gamma}e^{ins} + 1 + \sum_{n=-\infty}^{-1} e^{n\gamma}e^{ins} \\
&= \sum_{n=1}^{\infty} e^{-n\gamma}e^{ins} + 1 + \sum_{n=1}^{\infty} e^{-n\gamma}e^{-ins} \\
&= 1 + 2\sum_{n=1}^{\infty} e^{-n\gamma}\cos(ns),
\end{aligned}$$

the identity

$$1 + 2\sum_{n=1}^{\infty} e^{-n\gamma}\cos ns = \frac{\sinh(\gamma)}{\cosh(\gamma) - \cos(s)} \tag{9.2.22}$$

of Gradshteyn and Ryzhik [41, 1.461:2] implies that

$$\phi(s) = \frac{\sinh(\gamma)}{\cosh(\gamma) - \cos(s)}.$$

Consequently, with the choice $\gamma := \frac{\alpha^2}{4}$ in (9.2.21), that is, for

$$\phi(s) := \sum_{n=-\infty}^{\infty} e^{-|n|\frac{\alpha^2}{4}}e^{ins},$$

we find that

$$\phi(s) = \frac{\sinh(\frac{\alpha^2}{4})}{\cosh(\frac{\alpha^2}{4}) - \cos(s)}. \tag{9.2.23}$$

We will need two basic facts about the Fourier transform of 2π-periodic functions, see e.g., Katznelson [61, Sec. I]. If we denote the Fourier transform by $\hat{f}(n) := \frac{1}{2\pi} \int_{-\pi}^{\pi} f(s) e^{-ins}, \forall n$, the convolution theorem states that for periodic functions $f, g \in L^1[-\pi, \pi]$ that the convolution $(f \star g)(s) := \frac{1}{2\pi} \int_{-\pi}^{\pi} f(s-t) g(t) dt$ is a well-defined periodic function in $L^1[-\pi, \pi]$ and that $\widehat{(f \star g)}(n) = \hat{f}(n) \hat{g}(n), \forall n$. Moreover, for square-integrable 2π-periodic functions in $L^2[-\pi, \pi]$, the Parseval identity is $\sum_{n=-\infty}^{\infty} |\hat{f}(n)|^2 = \frac{1}{2\pi} \int_0^{2\pi} |f(s)|^2$.

Consequently, observing that $c_n = 0, n < -N, n > N$, the Parseval identity and the convolution formula imply that

$$
\begin{aligned}
\|y\|^2 &= \sum_{n=-N}^{N} e^{-\frac{|n|\alpha^2}{2}} |c_n|^2 \\
&= \left\| \left(e^{-\frac{|n|\alpha^2}{4}} c_n \right)_{n=-\infty}^{\infty} \right\|_{\ell^2}^2 \\
&= \left\| \left(\hat{\phi} \hat{y} \right)_{n=-\infty}^{\infty} \right\|_{\ell^2}^2 \\
&= \left\| \left(\widehat{\phi \star y} \right)_{n=-\infty}^{\infty} \right\|_{\ell^2}^2 \\
&= \| \phi \star y \|_{L^2[-\pi,\pi]}^2 \\
&= \int |\phi \star y|^2 \\
&= \int \left| \int \phi(s-t) y(t) dt \right|^2 ds \\
&= \int \left(\int \phi(s-t) y(t) dt \int \phi(s-t') y^*(t') dt' \right) ds \\
&= \int \int \phi(s-t) y(t) \phi(s-t') y^*(t') dt dt' ds \\
&= \int G(t, t') y(t) y^*(t') dt dt' ,
\end{aligned}
$$

that is,

$$
\|y\|^2 = \int G(t, t') y(t) y^*(t') dt dt',
$$

where

$$
G(t, t') := \int \phi(s-t) \phi(s-t') ds \tag{9.2.24}
$$

with

$$
\phi(s) = \frac{\sinh(\frac{\alpha^2}{4})}{\cosh(\frac{\alpha^2}{4}) - \cos(s)}. \tag{9.2.25}
$$

We can evaluate G using the identity (9.2.22) as follows: since

$$
\begin{aligned}
G(t, t') &= \int \phi(s-t) \phi(s-t') ds \\
&= \int \left(1 + 2 \sum_{n=1}^{\infty} e^{-n\frac{\alpha^2}{4}} \cos n(s-t) \right) \left(1 + 2 \sum_{n'=1}^{\infty} e^{-n'\frac{\alpha^2}{4}} \cos n'(s-t') \right) ds,
\end{aligned}
$$

and, for each product, we have

$$\int \cos n(s-t)\cos n'(s-t')ds$$

$$= \int (\cos ns \cos nt - \sin ns \sin nt)(\cos n's \cos n't' - \sin n's \sin n't')ds$$

$$= \delta_{n,n'}\int (\cos ns \cos nt - \sin ns \sin nt)(\cos ns \cos nt' - \sin ns \sin nt')ds\,.$$

Using the L^2-orthogonality of the cosines and the sines and the identities $\int \cos^2 ns = \pi$ and $\int \sin^2 ns = \pi$, we conclude that

$$\int (\cos ns \cos nt - \sin ns \sin nt)(\cos ns \cos nt' - \sin ns \sin nt')ds$$

$$= \pi(\cos nt \cos nt' + \sin nt \sin nt')$$

$$= \pi \cos n(t-t'),$$

and therefore

$$\int \cos n(s-t)\cos n'(s-t')ds = \pi\delta_{n,n'}\cos n(t-t')\,. \tag{9.2.26}$$

Consequently, we obtain

$$
\begin{aligned}
G(t,t') &= \int \left(1 + 2\sum_{n=1}^{\infty} e^{-n\frac{\alpha^2}{4}}\cos n(s-t)\right)\left(1 + 2\sum_{n'=1}^{\infty} e^{-n'\frac{\alpha^2}{4}}\cos n'(s-t')\right)ds \\
&= \int \left(1 + 4\sum_{n=1}^{\infty} e^{-n\frac{\alpha^2}{2}}\cos n(s-t)\cos n(s-t')\right)ds \\
&= 2\pi + 4\pi\sum_{n=1}^{\infty} e^{-n\frac{\alpha^2}{2}}\cos n(t-t'),
\end{aligned}
$$

and therefore, using the identity (9.2.22) again, we conclude

$$G(t,t') = 2\pi\frac{\sinh(\frac{\alpha^2}{2})}{\cosh(\frac{\alpha^2}{2}) - \cos(t-t')}\,.$$

Bibliography

[1] V. Adam, J. Hensman, and M. Sahani. Scalable transformed additive signal decomposition by non-conjugate Gaussian process inference. In *2016 IEEE 26th International Workshop on Machine Learning for Signal Processing (MLSP)*, pages 1–6. IEEE, 2016.

[2] M. Alvarez and N. D. Lawrence. Sparse convolved Gaussian processes for multi-output regression. In *Advances in Neural Information Processing Systems*, pages 57–64, 2009.

[3] M. A. Álvarez and N. D. Lawrence. Computationally efficient convolved multiple output Gaussian processes. *Journal of Machine Learning Research*, 12(May):1459–1500, 2011.

[4] M. A. Alvarez, L. Rosasco, and N. D. Lawrence. Kernels for vector-valued functions: a review. *Foundations and Trends® in Machine Learning*, 4(3):195–266, 2012.

[5] F. Auger, P. Flandrin, Y.-T. Lin, S. McLaughlin, S. Meignen, T. Oberlin, and H.-T. Wu. Time-frequency reassignment and synchrosqueezing: An overview. *IEEE Signal Processing Magazine*, 30(6):32–41, 2013.

[6] F. Bachoc. Cross validation and maximum likelihood estimations of hyper-parameters of gaussian processes with model misspecification. *Computational Statistics & Data Analysis*, 66:55–69, 2013.

[7] J. C. A. Barata and M. S. Hussein. The Moore–Penrose pseudoinverse: A tutorial review of the theory. *Brazilian Journal of Physics*, 42(1-2):146–165, 2012.

[8] B. Boashash. Estimating and interpreting the instantaneous frequency of a signal. I. Fundamentals. *Proceedings of the IEEE*, 80(4):520–538, 1992.

[9] S. Boyd, N. Parikh, E. Chu, B. Peleato, and J. Eckstein. Distributed optimization and statistical learning via the alternating direction method of multipliers. *Foundations and Trends® in Machine learning*, 3(1):1–122, 2011.

© The Author(s), under exclusive license to Springer Nature Switzerland AG 2021
H. Owhadi et al., *Kernel Mode Decomposition and the Programming of Kernels*, Surveys and Tutorials in the Applied Mathematical Sciences 8,
https://doi.org/10.1007/978-3-030-82171-5

[10] P Boyle and M. Frean. Dependent Gaussian processes. In *Advances in Neural Information Processing Systems*, pages 217–224, 2005.

[11] P. Boyle and M. Frean. Multiple output Gaussian process regression. 2005.

[12] Y. Chen, B. Hosseini, H. Owhadi, and A. M. Stuart. Solving and learning nonlinear pdes with gaussian processes. *arXiv preprint arXiv:2103.12959*, 2021.

[13] Y. Chen, H. Owhadi, and A. M. Stuart. Consistency of empirical Bayes and kernel flow for hierarchical parameter estimation. *Mathematics of Computation*, 2020. arXiv preprint arXiv:2005.11375.

[14] M. Costa, A. A. Priplata, L. A. Lipsitz, Z. Wu, N. E. Huang, A. L. Goldberger, and C.-K. Peng. Noise and poise: enhancement of postural complexity in the elderly with a stochastic-resonance–based therapy. *EPL (Europhysics Letters)*, 77(6):68008, 2007.

[15] K. T. Coughlin and K.-K. Tung. 11-year solar cycle in the stratosphere extracted by the empirical mode decomposition method. *Advances in Space Research*, 34(2):323–329, 2004.

[16] N. Cressie. The origins of Kriging. *Mathematical Geology*, 22(3):239–252, 1990.

[17] L. Csató. *Gaussian Processes: Iterative Sparse Approximations*. PhD thesis, Aston University Birmingham, UK, 2002.

[18] L. Csató and M. Opper. Sparse on-line Gaussian processes. *Neural Computation*, 14(3):641–668, 2002.

[19] L. Csató, M. Opper, and O. Winther. TAP Gibbs free energy, belief propagation and sparsity. In *Advances in Neural Information Processing Systems*, pages 657–663, 2002.

[20] D. A. T. Cummings, R. A. Irizarry, N. E. Huang, T. P. Endy, A. Nisalak, K. Ungchusak, and D. S. Burke. Travelling waves in the occurrence of dengue haemorrhagic fever in Thailand. *Nature*, 427(6972):344–347, 2004.

[21] I. Daubechies, J. Lu, and H.-T. Wu. Synchrosqueezed wavelet transforms: An empirical mode decomposition-like tool. *Applied and Computational Harmonic Analysis*, 30(2):243–261, 2011.

[22] I. Daubechies and S. Maes. A nonlinear squeezing of the continuous wavelet transform based on auditory nerve models. In A. Aldroubi and M. Unser, editors, *Wavelets in Medicine and Biology*, pages 527–546. World Scientific, 1996.

[23] R. Djemili, H. Bourouba, and M. C. Ammara Korba. Application of empirical mode decomposition and artificial neural network for the classification of normal and epileptic EEG signals. *Biocybernetics and Biomedical Engineering*, 36(1):285–291, 2016.

[24] K. Dragomiretskiy and D. Zosso. Variational mode decomposition. *IEEE Transactions on Signal Processing*, 62(3):531–544, 2014.

[25] R. M. Dudley. *Real Analysis and Probability*, volume 74 of *Cambridge Studies in Advanced Mathematics*. Cambridge University Press, Cambridge, 2002. Revised reprint of the 1989 original.

[26] N. Durrande, D. Ginsbourger, and O. Roustant. Additive kernels for Gaussian process modeling. *Annales de la Facultée de Sciences de Toulouse*, page 17, 2010.

[27] N. Durrande, D. Ginsbourger, and O. Roustant. Additive covariance kernels for high-dimensional Gaussian process modeling. In *Annales de la Faculté des sciences de Toulouse: Mathématiques*, volume 21, pages 481–499, 2012.

[28] N. Durrande, J. Hensman, M. Rattray, and N. D. Lawrence. Detecting periodicities with Gaussian processes. *PeerJ Computer Science*, 2:e50, 2016.

[29] N. Durrande, J. Hensman, M. Rattray, and N. D. Lawrence. Gaussian process models for periodicity detection. *PeerJ Computer Science*, 2016.

[30] D. K. Duvenaud, H. Nickisch, and C. E. Rasmussen. Additive Gaussian processes. In *Advances in Neural Information Processing Systems*, pages 226–234, 2011.

[31] H. W. Engl, M. Hanke, and A. Neubauer. *Regularization of Inverse Problems*, volume 375. Springer Science & Business Media, 1996.

[32] Y. Fan, G. M. James, and P. Radchenko. Functional additive regression. *The Annals of Statistics*, 43(5):2296–2325, 2015.

[33] M. Feldman. Time-varying vibration decomposition and analysis based on the Hilbert transform. *Journal of Sound and Vibration*, 295(3-5):518–530, 2006.

[34] P. Flandrin and P. Goncalves. Empirical mode decompositions as data-driven wavelet-like expansions. *International Journal of Wavelets, Multiresolution and Information Processing*, 2(04):477–496, 2004.

[35] P. Flandrin, P. Gonçalves, and G. Rilling. EMD equivalent filter banks, from interpretation to applications. In *Hilbert-Huang Transform and its Applications*, pages 57–74. World Scientific, 2005.

[36] P. Flandrin, G. Rilling, and P. Goncalves. Empirical mode decomposition as a filter bank. *IEEE Signal Processing Letters*, 11(2):112–114, 2004.

[37] D. Gabor. Theory of communication. part 1: The analysis of information. *Journal of the Institution of Electrical Engineers-Part III: Radio and Communication Engineering*, 93(26):429–441, 1946.

[38] L. S. Gandin. Objective analysis of meteorological fields: Gidrometeorotogicheskoizeda- tel'stvo(GIMIZ), Leningrad (translated by Israel Program for Scientific Translations, Jerusalem, 1965, 238 pp.). 1963.

[39] J. Gilles. Empirical wavelet transform. *IEEE Transactions on Signal Processing*, 61(16):3999–4010, 2013.

[40] I. Goodfellow, Y. Bengio, A. Courville, and Y. Bengio. *Deep learning*, volume 1. MIT press Cambridge, 2016.

[41] I. S. Gradshteyn and I. M. Ryzhik. *Table of Integrals, Series, and Products*. Academic, 6th edition, 2000.

[42] B. Hamzi, R. Maulik, and H. Owhadi. Data-driven geophysical forecasting: Simple, low-cost, and accurate baselines with kernel methods. *arXiv preprint arXiv:2103.10935*, 2021.

[43] B. Hamzi and H. Owhadi. Learning dynamical systems from data: a simple cross-validation perspective. *arXiv:2007.05074*, 2020.

[44] B. Hamzi and H. Owhadi. Learning dynamical systems from data: A simple cross-validation perspective, part i: Parametric kernel flows. *Physica D: Nonlinear Phenomena*, 421:132817, 2021.

[45] T. Hastie and R. Tibshirani. Generalized additive models. *Statistical Science*, 1(w3):297–310, 1986.

[46] T. J. Hastie and R. J. Tibshirani. *Generalized Additive Models*, volume 43. CRC press, 1990.

[47] K. He, X. Zhang, S. Ren, and J. Sun. Deep residual learning for image recognition. In *Proceedings of the IEEE conference on computer vision and pattern recognition*, pages 770–778, 2016.

[48] S. Helgason. *The Radon Transform*, volume 2. Springer, 1999.

[49] J. Hensman, N. Durrande, and A. Solin. Variational Fourier features for Gaussian processes. *The Journal of Machine Learning Research*, 18(1):5537–5588, 2017.

[50] J. Hensman, N. Fusi, and N. D. Lawrence. Gaussian processes for big data. In *Uncertainty in Artificial Intelligence*, page 282. Citeseer, 2013.

[51] J. Hensman, A. G. Matthews, and Z. Ghahramani. Scalable variational Gaussian process classification. *Proceedings of Machine Learning Research*, 38:351–360, 2015.

[52] M. D. Hoffman, D. M. Blei, C. Wang, and J. Paisley. Stochastic variational inference. *The Journal of Machine Learning Research*, 14(1):1303–1347, 2013.

[53] T. Y. Hou and Z. Shi. Adaptive data analysis via sparse time-frequency representation. *Advances in Adaptive Data Analysis*, 3(01n02):1–28, 2011.

[54] T. Y. Hou, Z. Shi, and P. Tavallali. Sparse time frequency representations and dynamical systems. *Communications in Mathematical Sciences*, 13(3):673–694, 2015.

[55] C. Huang, L. Yang, and Y. Wang. Convergence of a convolution-filtering-based algorithm for empirical mode decomposition. *Advances in Adaptive Data Analysis*, 1(04):561–571, 2009.

[56] N. E. Huang. Introduction to the Hilbert-Huang transform and its related mathematical problems. In N. E. Huang and S. S. P. Shen, editors, *Hilbert-Huang Transformation and it Applications*, pages 1–26. World Scientific, 2014.

[57] N. E. Huang and S. S. P. Shen. *Hilbert-Huang Transform and its Applications*, volume 16. World Scientific, 2014.

[58] N. E. Huang, Z. Shen, S. R. Long, M. C. Wu, H. H. Shih, Q. Zheng, N.-C. Yen, C. C. Tung, and H. H. Liu. The empirical mode decomposition and the Hilbert spectrum for nonlinear and non-stationary time series analysis. *Proceedings of the Royal Society of London. Series A: Mathematical, Physical and Engineering Sciences*, 454(1971):903–995, 1998.

[59] N. E. Huang and Z. Wu. A review on Hilbert-Huang transform: Method and its applications to geophysical studies. *Reviews of Geophysics*, 46(2), 2008.

[60] M. Hutson. Has artificial intelligence become alchemy? *Science*, 360(6388):861, 2018.

[61] Y. Katznelson. *An Introduction to Harmonic Analysis*. Cambridge University Press, 2004.

[62] R. Kress. *Linear Integral Equations*, volume 82. Springer, 3rd edition, 1989.

[63] N. D. Lawrence, M. Seeger, and R. Herbrich. Fast sparse Gaussian process methods: The informative vector machine. In *Advances in Neural Information Processing Systems*, pages 625–632, 2003.

[64] Y. LeCun. The deep learning- applied math connection. In *SIAM Conference on Mathematics of Data Science (MDS20)*, 2020. Streamd live on June 24, 2020, https://www.youtube.com/watch?v=y9gutjsvc1c& feature=youtu.be&t=676.

[65] Y. LeCun, Y. Bengio, and G. Hinton. Deep learning. *Nature*, 521(7553):436–444, 2015.

[66] C. Li and M. Liang. Time–frequency signal analysis for gearbox fault diagnosis using a generalized synchrosqueezing transform. *Mechanical Systems and Signal Processing*, 26:205–217, 2012.

[67] C.-Y. Lin, L. Su, and H.-T. Wu. Wave-shape function analysis. *Journal of Fourier Analysis and Applications*, 24(2):451–505, 2018.

[68] L. Lin, Y. Wang, and H. Zhou. Iterative filtering as an alternative algorithm for empirical mode decomposition. *Advances in Adaptive Data Analysis*, 1(04):543–560, 2009.

[69] A. Liutkus, R. Badeau, and G. Richard. Gaussian processes for underdetermined source separation. *IEEE Transactions on Signal Processing*, 59(7):3155–3167, 2011.

[70] M. Lukić and J. Beder. Stochastic processes with sample paths in reproducing kernel Hilbert spaces. *Transactions of the American Mathematical Society*, 353(10):3945–3969, 2001.

[71] W. Ma, S. Yin, C. Jiang, and Y. Zhang. Variational mode decomposition denoising combined with the Hausdorff distance. *Review of Scientific Instruments*, 88(3):035109, 2017.

[72] S. Maji, A. C. Berg, and J. Malik. Efficient classification for additive kernel SVMs. *IEEE Transactions on Pattern Analysis and Machine Intelligence*, 35:66–77, 2013.

[73] G. Matheron. Principles of geostatistics. *Economic Geology*, 58(8):1246–1266, 1963.

[74] G. Matheron. *Traité de Géostatistique Appliquée. 2. Le Krigeage*. Editions Technip, 1963.

[75] G. J. McLachlan, S. X. Lee, and S. I. Rathnayake. Finite mixture models. *Annual Review of Statistics and its Application*, 6:355–378, 2019.

[76] A. Melkumyan and F. Ramos. Multi-kernel Gaussian processes. In *Twenty-second International Joint Conference on Artificial Intelligence*, 2011.

[77] R. K. Merton. *The Sociology of Science: Theoretical and Empirical Investigations*. University of Chicago Press, 1973.

[78] C. A. Micchelli and T. J. Rivlin. A survey of optimal recovery. In *Optimal Estimation in Approximation Theory*, pages 1–54. Springer, 1977.

[79] T. Oberlin, S. Meignen, and V. Perrier. The Fourier-based synchrosqueezing transform. In *2014 IEEE International Conference on Acoustics, Speech and Signal Processing (ICASSP)*, pages 315–319. IEEE, 2014.

[80] H. Owhadi. Multigrid with rough coefficients and multiresolution operator decomposition from hierarchical information games. *SIAM Review*, 59(1):99–149, 2017.

[81] H. Owhadi. Do ideas have shape? plato's theory of forms as the continuous limit of artificial neural networks. *arXiv preprint arXiv:2008.03920*, 2020.

[82] H. Owhadi and C. Scovel. *Operator Adapted Wavelets, Fast Solvers, and Numerical Homogenization, from a game theoretic approach to numerical approximation and algorithm design*. Cambridge Monographs on Applied and Computational Mathematics. Cambridge University Press, 2019.

[83] H. Owhadi, C. Scovel, and F. Schäfer. Statistical numerical approximation. *Notices of the AMS*, 66(10):1608–1617, 2019.

[84] H. Owhadi and G. R. Yoo. Kernel flows: From learning kernels from data into the abyss. *Journal of Computational Physics*, 389:22–47, 2019.

[85] S. Park and S. Choi. Gaussian processes for source separation. In *2008 IEEE International Conference on Acoustics, Speech and Signal Processing*, pages 1909–1912. IEEE, 2008.

[86] A. B. Patel, M. T. Nguyen, and R. Baraniuk. A probabilistic framework for deep learning. In *Advances in Neural Information Processing Systems*, pages 2558–2566, 2016.

[87] T. A. Plate. Accuracy versus interpretability in flexible modeling: Implementing a tradeoff using Gaussian process models. *Behaviormetrika*, 26(1):29–50, 1999.

[88] D. Preoţiuc-Pietro and T. Cohn. A temporal model of text periodicities using Gaussian processes. In *Proceedings of the 2013 Conference on Empirical Methods in Natural Language Processing*, pages 977–988, 2013.

[89] J. Quiñonero-Candela. *Learning with Uncertainty: Gaussian Processes and Relevance Vector Machines*. PhD thesis, Technical University of Denmark Lyngby, Denmark, 2004.

[90] J. Quiñonero-Candela and C. E. Rasmussen. A unifying view of sparse approximate Gaussian process regression. *Journal of Machine Learning Research*, 6(Dec):1939–1959, 2005.

[91]   M. Raissi, P. Perdikaris, and G. E. Karniadakis. Machine learning of linear differential equations using Gaussian processes. *Journal of Computational Physics*, 348:683–693, 2017.

[92]   M. M. Rao. *Foundations of Stochastic Analysis*. Academic Press, 1981.

[93]   C. E. Rasmussen. Gaussian processes in machine learning. In *Summer School on Machine Learning*, pages 63–71. Springer, 2003.

[94]   C. E. Rasmussen and C. K. I. Williams. *Gaussian Processes for Machine Learning*, volume 2. MIT press Cambridge, MA, 2006.

[95]   G. Rilling and P. Flandrin. One or two frequencies? The empirical mode decomposition answers. *IEEE Transactions on Signal Processing*, 56(1):85–95, 2007.

[96]   G. Rilling, P. Flandrin, and P. Goncalves. On empirical mode decomposition and its algorithms. In *IEEE-EURASIP Workshop on Nonlinear Signal and Image Processing*, volume 3, pages 8–11. NSIP-03, Grado (I), 2003.

[97]   F. Schäfer, T. J. Sullivan, and H. Owhadi. Compression, inversion, and approximate PCA of dense kernel matrices at near-linear computational complexity. *Multiscale Modeling & Simulation*, SIAM, 19(2):688–730, 2021.

[98]   B. Schölkopf, A. J. Smola, F. Bach, et al. *Learning with Kernels: Support Vector Machines, Regularization, Optimization, and Beyond*. MIT press, 2002.

[99]   A. Schwaighofer and V. Tresp. Transductive and inductive methods for approximate Gaussian process regression. In *Advances in Neural Information Processing Systems*, pages 977–984, 2003.

[100] M. Seeger. Bayesian Gaussian process models: PAC-Bayesian generalisation error bounds and sparse approximations. Technical report, University of Edinburgh, 2003.

[101] M. Seeger, C. K. I. Williams, and N. D. Lawrence. Fast forward selection to speed up sparse Gaussian process regression. In *Proceedings of the Ninth International Workshop on Artificial Intelligence and Statistics*, 2003.

[102] J. Shawe-Taylor, N. Cristianini, et al. *Kernel methods for pattern analysis*. Cambridge university press, 2004.

[103] A. J. Smola and P. L. Bartlett. Sparse greedy Gaussian process regression. In *Advances in Neural Information Processing Systems*, pages 619–625, 2001.

[104] E. Snelson and Z. Ghahramani. Sparse Gaussian processes using pseudo-inputs. In *Advances in Neural Information Processing Systems*, pages 1257–1264, 2006.

[105] E. P. Souza Neto, M. A. Custaud, J. C. Cejka, P. Abry, J. Frutoso, C. Gharib, and P. Flandrin. Assessment of cardiovascular autonomic control by the empirical mode decomposition. *Methods of Information in Medicine*, 43(01):60–65, 2004.

[106] M. L. Stein. A comparison of generalized cross validation and modified maximum likelihood for estimating the parameters of a stochastic process. *The Annals of Statistics*, pages 1139–1157, 1990.

[107] I. Steinwart and A. Christmann. *Support Vector Machines*. Springer Science & Business Media, 2008.

[108] S. M. Stigler. Stigler's law of eponymy. *Transactions of the New York Academy of Sciences*, 39(1 Series II):147–157, 1980.

[109] C. J. Stone. Additive regression and other nonparametric models. *The Annals of Statistics*, pages 689–705, 1985.

[110] G. Thakur. The synchrosqueezing transform for instantaneous spectral analysis. In *Excursions in Harmonic Analysis, Volume 4*, pages 397–406. Springer, 2015.

[111] G. Thakur, E. Brevdo, N. S. Fučkar, and H.-T. Wu. The synchrosqueezing algorithm for time-varying spectral analysis: Robustness properties and new paleoclimate applications. *Signal Processing*, 93(5):1079–1094, 2013.

[112] M. Titsias. Variational learning of inducing variables in sparse Gaussian processes. In *Artificial Intelligence and Statistics*, pages 567–574, 2009.

[113] M. K. Titsias and M. Lázaro-Gredilla. Spike and slab variational inference for multi-task and multiple kernel learning. In *Advances in Neural Information Processing Systems*, pages 2339–2347, 2011.

[114] M. E. Torres, M. A. Colominas, G. Schlotthauer, and P. Flandrin. A complete ensemble empirical mode decomposition with adaptive noise. In *2011 IEEE International Conference on Acoustics, Speech and Signal Processing (ICASSP)*, pages 4144–4147. IEEE, 2011.

[115] V. Tresp. A Bayesian committee machine. *Neural computation*, 12(11):2719–2741, 2000.

[116] S. Wang, X. Chen, G. Cai, B. Chen, X. Li, and Z. He. Matching demodulation transform and synchrosqueezing in time-frequency analysis. *IEEE Transactions on Signal Processing*, 62(1):69–84, 2013.

[117] C. K. I. Williams and M. Seeger. Using the Nyström method to speed up kernel machines. In *Advances in Neural Information Processing Systems*, pages 682–688, 2001.

[118] K. I. Williams, C and C. E. Rasmussen. Gaussian processes for regression. In *Advances in Neural Information Processing Systems*, pages 514–520, 1996.

[119] Z. Wu and N. E. Huang. A study of the characteristics of white noise using the empirical mode decomposition method. *Proceedings of the Royal Society of London. Series A: Mathematical, Physical and Engineering Sciences*, 460(2046):1597–1611, 2004.

[120] Z. Wu and N. E. Huang. Ensemble empirical mode decomposition: a noise-assisted data analysis method. *Advances in Adaptive Data Analysis*, 1(01):1–41, 2009.

[121] Z. Wu, E. K. Schnieder, Z.-Z. Hu, and L. Cao. *The Impact of Global Warming on ENSO Variability in Climate Records*, volume 110. Center for Ocean-Land-Atmosphere Studies Calverton, 2001.

[122] T. W. Yee. *Vector Generalized Linear and Additive Models: with an Implementation in R*. Springer, 2015.

[123] T. W. Yee and C. J. Wild. Vector generalized additive models. *Journal of the Royal Statistical Society: Series B (Methodological)*, 58(3):481–493, 1996.

[124] G. R. Yoo. *Learning Patterns with Kernels and Learning Kernels from Patterns*. PhD thesis, California Institute of Technology, 2020.

[125] K. Yosida. *Functional Analysis*. Springer-Verlag, Berlin, 5th edition, 1978.

[126] K. Yu, V. Tresp, and A. Schwaighofer. Learning Gaussian processes from multiple tasks. In *Proceedings of the 22nd International Conference on Machine Learning*, pages 1012–1019, 2005.

Index

H. Owhadi et al., *Kernel Mode Decomposition and the Programming of
Kernels*, Surveys and Tutorials in the Applied Mathematical Sciences 8,
https://doi.org/10.1007/978-3-030-82171-5

Printed in the United States
by Baker & Taylor Publisher Services